Optical Fibre Devices

Series in Optics and Optoelectronics

Series Editors:

RGW Brown, University of Nottingham, UK
ER Pike, Kings College, London, UK

Other titles in the series

The Optical Transfer Function of Imaging Systems
TL Williams

Super-Radiance: Multiatomic Coherent Emission
MG Benedict, AM Ermolaev, UA Malyshev, IV Sokolov and ED Trifonov

Solar Cells and Optics for Photovoltaic Concentration
A Luque

Applications of Silicon–Germanium Heterostructure Devices
CK Maiti and GA Armstrong

Forthcoming titles in the series

Diode Lasers
D Sands

High Aperture Focussing of Electromagnetic Waves and Applications in Optical Microscopy
CJR Sheppard and P Torok

Power and Energy Handling Capabilities of Optical Materials, Components and Systems
RM Wood

The Practical Application of the Moiré Fringe Method
CA Walker (ed)

Transparent Conductive Coatings
CI Bright

Stimulated Brillouin Scattering: Theory and Applications
MJ Damzen, VI Vlad, V Babin and A Mocofanescu

Other titles of interest

Thin-Film Optical Filters (Third Edition)
H Angus Macleod

Series in Optics and Optoelectronics

Optical Fibre Devices

J-P Goure and
I Verrier

*TSI Laboratory,
Faculty of Science and Technology
University of Saint-Etienne*

Institute of Physics Publishing
Bristol and Philadelphia

© IOP Publishing Ltd 2002

All rights reserved. No part of this publication may be reproduced, stored in a retrieval system or transmitted in any form or by any means, electronic, mechanical, photocopying, recording or otherwise, without the prior permission of the publisher. Multiple copying is permitted in accordance with the terms of licences issued by the Copyright Licensing Agency under the terms of its agreement with the Committee of Vice-Chancellors and Principals.

British Library Cataloguing-in-Publication Data

A catalogue record for this book is available from the British Library.

ISBN 0 7503 0811 7

Library of Congress Cataloging-in-Publication Data are available

Commissioning Editor: Tom Spicer
Production Editor: Simon Laurenson
Production Control: Sarah Plenty
Cover Design: Victoria Le Billon
Marketing Executive: Laura Serratrice

Published by Institute of Physics Publishing, wholly owned by The Institute of Physics, London

Institute of Physics Publishing, Dirac House, Temple Back, Bristol BS1 6BE, UK

US Office: Institute of Physics Publishing, The Public Ledger Building, Suite 1035, 150 South Independence Mall West, Philadelphia, PA 19106, USA

Typeset in England by Alden Bookset
Printed in the UK by MPG Books Ltd, Bodmin, Cornwall

Contents

PREFACE AND ACKNOWLEDGMENTS	xi
INTRODUCTION	xiv
1. OPTICAL FIBRE AND PROPAGATION	**1**
1.1 Geometry and characteristics of optical fibres	**1**
1.1.1 Refraction and reflection of a light ray	2
1.1.2 Optical fibre	5
1.1.3 Step index multimode fibre	6
1.1.4 Graded index multimode fibre	6
1.1.5 Numerical aperture	8
1.1.6 Time delay	8
1.2 Propagation in optical fibres	**10**
1.2.1 Multimode fibres	10
1.2.2 Single-mode fibre	14
1.3 Characteristics and measurements	**15**
1.3.1 Numerical aperture	15
1.3.2 Diameter control and measurement	16
1.3.3 Attenuation measurement	16
1.3.3.1 Cut-fibre method	17
1.3.3.2 Retro-diffusion method	18
1.4 Birefringence	**19**
1.4.1 Shape birefringence	19
1.4.2 Stress birefringence	20
1.4.3 Stokes parameters and Poincaré sphere	20
1.5 Fabrication and materials	**22**

1.5.1 Fabrication	22
1.5.2 Materials	24
1.5.3 Rare-earth doped fibres	25

1.6 Light sources   25
     1.6.1 Laser diodes   26
     1.6.2 LED sources   28

References for Chapter 1   28

Exercises   30

2. COUPLING: MICROCOMPONENTS, TAPERS, SPLICES, CONNECTORS   31

2.1 Fibre ends   31

2.2 Microcomponents   33
     2.2.1 GRIN lenses   33
     2.2.2 Microlenses   35
     2.2.3 Tapers   39

2.3 Coupling from fibre to fibre   49
     2.3.1 Alignment losses, butt coupling   49
         2.3.1.1 Splice of circular fields   49
         2.3.1.2 Splice of elliptic fields   53
     2.3.2 Coupling with microcomponents   54
     2.3.3 Non-intrusive tap   56
     2.3.4 Connectors   58

2.4 Coupling from fibre to waveguide   58
     2.4.1 Butt end coupling   60
     2.4.2 Coupling using grooves   61
     2.4.3 Other methods   63

2.5 Coupling from semiconductor lasers or LED into fibres   64
     2.5.1 Multimode fibre   64
     2.5.2 Single-mode fibre   67

References for Chapter 2   69

Exercises   73

3. DEVICES BASED ON COUPLING EFFECT WITH NON-POLARIZED LIGHT — 75

3.1 Coupling theory for circular fibres — 76

3.2 Directional couplers — 80

    3.2.1 X coupler, 2×2 coupler — 80
    3.2.2 Y coupler — 89
    3.2.3 Star coupler — 90
    3.2.4 Micro-optics coupler — 93

3.3 Bragg gratings — 93

    3.3.1 Production — 93
    3.3.2 Theory — 96

3.4 Wavelength multiplexers and demultiplexers — 99

    3.4.1 With external components — 99
    3.4.2 All-optical-fibre devices — 101
        3.4.1.1 Using Bragg grating devices — 101
        3.4.1.2 Using mechanical devices — 102

3.5 Frequency and phase shifters — 103

    3.5.1 Frequency shifters — 103
    3.5.2 Phase shifters — 105

3.6 Wavelength and modal filters — 106

    3.6.1 Wavelength filters — 106
    3.6.2 Filters based on modal filtering — 110

3.7 Linear switches and taps — 110

    3.7.1 Switches — 110
    3.7.2 Taps — 114

3.8 Modulators — 115

    3.8.1 Phase modulators — 115
    3.8.2 Intensity modulators — 116

3.9 Loops and rings — 116

    3.9.1 Resonators — 117
    3.9.2 Delay lines and circulators — 117

viii   Contents

3.10 Interferometers	118
References for Chapter 3	122
Exercises	128
4. DEVICES USING POLARIZED LIGHT	**129**
4.1 Polarization in single-mode fibres	129
4.2 X-couplers using birefringent fibres	134
4.2.1 Polarization-maintaining couplers	134
4.2.2 Polarization splitting and polarization-selective couplers	135
4.3 Birefringent fibre polarization coupler	137
4.4 Polarization devices	140
4.4.1 Polarizers	140
4.4.1.1 Linear polarizer using birefringent material	141
4.4.1.2 Linear polarizer using thin metal films	142
4.4.1.3 Linear polarizer using multiple interfaces	146
4.4.1.4 Circular polarizers	150
4.4.2 Depolarizers	151
4.4.3 Polarization state controllers	153
4.4.3.1 Polarization controller using stressed fibres	153
4.4.3.2 Polarization controller using intrinsic fibre birefringence	156
4.4.3.3 Polarization controller using liquid crystals	158
4.4.4 Isolators	159
4.5 Polarimeters – Interferometers	160
References for Chapter 4	162
Exercises	164
5. DEVICES USING NON-LINEARITIES IN FIBRES	**165**
5.1 Devices based on stimulated Raman scattering (SRS)	167
5.1.1 Basic principle	167
5.1.2 Amplification based on Raman effect	172
5.1.3 Raman gain in a re-entrant fibre loop	177

5.2 Devices based on stimulated Brillouin scattering (SBS)   179

    5.2.1 Basic principle   179
    5.2.2 Fibre Brillouin amplifiers   180
    5.2.3 Brillouin laser based on a fibre ring resonator   183

5.3 Parametric four wave mixing   186

5.4 Kerr non-linearities in optical fibres – solitons   188

    5.4.1 Basic principle   188
    5.4.2 Optical pulse compression   191
    5.4.3 Soliton phenomenon   191
    5.4.4 The soliton laser   195

5.5 Switches   196

    5.5.1 Soliton switching in non-linear directional couplers   196
    5.5.2 Switches using non-linear couplers   199
    5.5.3 Switching using birefringent fibres   202
       5.5.3.1 High birefringent fibres   202
       5.5.3.2 Low birefringent fibres   204
       5.5.3.3 Twisted birefringent fibre   206
       5.5.3.4 Birefringent fibres with cross axis   207
    5.5.4 Switching using non-linear fibre loop mirror   210
       5.5.4.1 With non-birefringent fibre   210
       5.5.4.2 With PANDA fibre   210

5.6 Non-linear fibre interferometer   211

5.7 Modulator and logic gate   212

5.8 Optical fibre transistor   214

References for Chapter 5   215

Exercises   219

6. LASERS AND AMPLIFIERS BASED ON RARE-EARTH DOPED FIBRES   221

6.1 Rare-earth doped fibre amplifiers (REDFAs)   222

    6.1.1 Principle   223
    6.1.2 Example: Er^{3+} doped optical fibre amplifiers   225
       6.1.2.1 Energy levels   225
       6.1.2.2 Approximate models of the levels system   227
       6.1.2.3 Optical gain   232
       6.1.2.4 Noise   233

    6.1.3 Other doped optical fibres   235
    6.1.4 Device aspects of fibre amplifiers, performance and applications   236

6.2 Rare-earth doped fibre lasers (REDFLs)   238

    6.2.1 Principle   238
    6.2.2 Fibre laser using gratings   239
    6.2.3 Fibre laser using directional couplers   241
    6.2.4 Fibre laser using fibre reflectors   243
    6.2.5 Q-switching fibre laser   244
    6.2.6 Mode-locking fibre laser   245
    6.2.7 Tunable operation   246
    6.2.8 Superfluorescent rare-earth doped fibre sources   247

References for Chapter 6   **247**

Exercises   **252**

7 RECENT DEVELOPMENTS AND CONCLUSION   253

7.1 Synthesis of devices described in this book   253

7.2 New developments   257

References for Chapter 7   **259**

List of symbols   **260**

Solutions to exercises   **263**

INDEX   **266**

Preface and Acknowledgments

The rapid acceptance of fibre optics technology in commercial systems applications has brought forward an increased need for guided wave optical components (passive and active) which permit coupling, multiplexing and demultiplexing, switching and distributing of light signals propagating in optical waveguides. Optical components such as light sources fibre couplers, splices, connectors, directional couplers, multiplexers, switches and modulators are essential elements for transmission and sensing systems. Splices and connectors between fibres are important because splice losses remain a factor of primary importance in existing and future fibre optics network design.

In the search for simplicity, all-fibre components associated with optical functions have recently received considerable attention, mainly because of the strong potential benefits such components may play in the field of light wave technology, including optical fibre communications, opto-electronic signal processing and optical fibre sensing. Devices constructed from optical fibres may have advantages such as low insertion loss due to circular symmetry.

Our purpose in this book is to provide an overview, focused on those all-fibre devices that are already available or have the greatest potential for providing improved performance. We have not attempted to catalogue all the variations that have been proposed, but we describe their physical principles and the results, and we give some applications. It should be emphasized, however, that most of the components and devices are still in the laboratory stage.

We have also tried to provide references to guide the interested reader to more detailed studies, but we have not attempted to quote all the technical contributions that have advanced the field to its present stage.

We wish to thank our colleagues at TSI Laboratory (UMR CNRS 5516) for their support and would like also to express our sincere gratitude to Colette Veillas, Marie Laure Pugnière, Valérie Plomb and Jeanine Percet for the drawings and typing.

J-P Goure
I Verrier

Introduction

The new technology of optical low-loss transmission has the advantages of large information carrying capacity, immunity from electromagnetic interference and small size and weight. The optical fibre has become the medium for communications not only for transoceanic communication, but also for local industrial networks.

Fibre optics systems represent the largest growing market in the electro-optic industry. The major fibre optics applications include telephones, cable TV, industrial automation, sensors and computers.

Many textbooks and monographs on optical fibres have been published throughout the development of low loss optical fibres. Most of these have been devoted only to the theoretical aspects of optical waveguides. The book describes major components, including aspects of linear and non-linear optics. It discusses in detail optical fibre devices in the present state of development.

Likewise, micro-optic components and integrated optical components, which are based on forming patterns of optical waveguides on the surface of a planar substrate, are not covered here, as reviews of recent development in the fabrication and performance of these optical components have been published elsewhere.

Chapter 1 is devoted to a description of optical fibres and their characterization. We also give some explanation of the methods used in the realization of these devices. In the second chapter we examine the problem of splices and connectors.

The more usual devices using linear effects are described in Chapter 3 and devices using polarization effects in Chapter 4.

In addition, optical fibres can be utilized not only for their transmission characteristics, but also as non-linear devices. A significant aspect described in Chapter 5 is the use of non-linear optical effects in order to provide special useful optical device functions such as optical gating and switching, optical pulse shaping and short pulse generation, which may find interesting new applications in future optical signal transmission, signal processing or optical sensing systems.

Fibre optical amplification and lasers using doped fibres are treated in Chapter 6.

Finally, Chapter 7 is devoted to the synthesis of the most significant devices described in this book and to recent developments promising important future applications.

Chapter 1

OPTICAL FIBRE AND PROPAGATION

To understand the way optical fibre devices work, it is necessary to study some basic concepts of optics and physics. This chapter is intended to introduce these ideas to those who may not have studied the field, and to review them for those who have. The reader is urged to seek additional information in the references at the end of this Chapter (Jeunhomme 1988, Miller *et al.* 1986, Miller and Kaminov 1988, Neumann 1988, Marcuse 1974, Adams 1981, Snyder and Love 1983, Vassallo 1991, Kapany and Burke 1972, Senior 1992). Reviews of the development, fabrication and performance of optical components (Stolen and De Paula 1987, Goure *et al.* 1989, Noda and Yokohama 1988, Dakin and Culshaw 1988, Dakin and Culshaw 1989) and sensing systems (Culshaw 1984, Jones 1987) have been published.

1.1 GEOMETRY AND CHARACTERISTICS OF OPTICAL FIBRES

The visible light by which we see the world around us is part of the range of electromagnetic waves, which extends from radio frequencies to high-energy gamma radiation. These waves are formed by electric (**E**) and magnetic (**H**) fields, which propagate through a vacuum or a dielectric medium with a wave vector **k** (see Figure 1.1). They have as their most distinguishing features their frequencies v of oscillation. The light speed c, the frequencies of oscillation v and wavelengths in empty space λ are connected by:

$$\lambda = c/v \qquad (1.1)$$

$c = 3 \times 10^8$ ms^{-1}. The wave vector is defined by:

$$\mathbf{k} = (2\pi/\lambda)\mathbf{e}_3 \qquad (1.2)$$

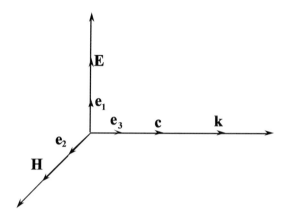

Figure 1.1. Scheme of an electromagnetic wave in a homogeneous medium: **E** electric vector; **H** magnetic vector; **c** speed in empty space; **k** wave vector.

where e_3 is the unit vector in the direction of propagation. The wavelength range for visible light is from 400 nanometres (nm) to about 700 nm (1 nm = 10^{-9}m). In the technology optical fibres, the most useful sources of electromagnetic radiation operate in the visible or in near infrared region at wavelengths around 800 and 1500 nm.

1.1.1 Refraction and reflection of a light ray

In most simple cases, the results of the interaction of an electromagnetic wave with an isotropic and homogeneous medium can be expressed in terms of a single number n, the refractive index of the medium. The refractive index is the ratio of light speed c in empty space and the light speed v in the medium.

$$n = c/v \qquad (1.3)$$

The refractive index is always greater than 1, since the speed of light in a medium is always less than in empty space. In glass, it varies from about 1.4 to about 1.9, in gallium arsenide (GaAs) it varies from 3.5 to 3.7.

In an isotropic homogeneous medium, the refractive index is constant in space and light travels in a straight line. When the light meets a variation or discontinuity in the refractive index, the light rays will be bent from their initial direction.

Where the refractive index within a material varies, the behaviour of the light is governed by the index change in space. If the change of refractive index is abrupt, as in the boundary, between air and glass, the change in direction is governed by the laws of geometrical optics

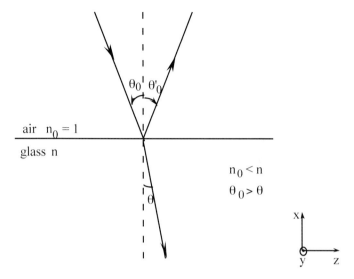

Figure 1.2. Snell's law. Reflection and refraction of light at air–glass interface.

(see Figure 1.2). If θ_0 is the angle of incidence between a ray and a line perpendicular to the interface at the point where the light ray strikes the interface, then:

- the three rays (incident, reflected, refracted) are in a same plane,
- the angle of reflection θ'_0 also measured with respect to the same perpendicular, is equal to the angle of incidence θ_0 : $|\theta_0| = |\theta'_0|$
- the angle θ of the transmitted light is given by the relation:

$$n_0 \sin \theta_0 = n \sin \theta \qquad (1.4)$$

The second of these propositions is known as the law of reflection; the third as the law of refraction or Snell's Law.

If now rays in glass (index n) are incident on the interface at a range of angles (see Figure 1.3), ray 1 is refracted into air. Ray 2 is incident at an angle such that the refracted angle in air is 90°. If the angle of incidence of ray 3 is larger than θ_2, all light is reflected back into the incident medium, and no light enters the second medium. Nevertheless, in this last case, the electromagnetic field does not entirely vanish in the second medium. There is in fact a small flow of energy across the boundary. This energy induces a non-homogeneous wave which falls off exponentially with the distance x in the second medium from its boundary.

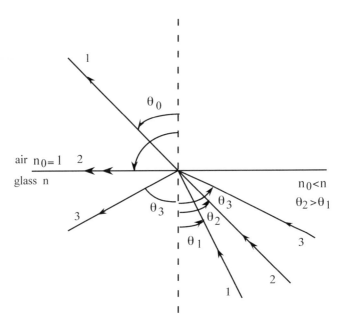

Figure 1.3. Illustration of total internal reflection for ray 3 (ray 1 is transmitted).

There is thus a penetration depth of the wave which vanishes very rapidly and is called an *evanescent wave*. Although there is a wave in the second medium there is no energy corresponding to it. (This can be verified by calculating the Poynting vector in the second medium: its average is zero). There is no light actually transmitted into the second medium. The light is said to be *totally internally reflected*. For all angles of incidence greater than this *critical angle*, total internal reflection will occur. This critical angle occurs for the angle of incidence at which the transmitted ray is refracted along the surface of the interface (the case illustrated by ray 2). Setting the transmission angle equal to 90°, the critical angle θ_{ca} corresponding to θ_2, is found from the relationship:

$$\sin \theta_{ca} = n_0/n \tag{1.5}$$

To understand the behaviour of light rays at the interface, let us talk in terms of wave propagation. The incident ray is associated with an electromagnetic wave of three vector components ($\mathbf{E_0}$, $\mathbf{H_0}$, $\mathbf{k_0}$). $\mathbf{k_0}$ is the propagation vector in the medium of refractive index n_0 and the lines of force corresponding to $\mathbf{k_0}$ are represented by the ray $\{|\mathbf{k_0}| = 2\pi n_0/\lambda = \omega/v_0\}$. $\mathbf{E_0}$ and $\mathbf{H_0}$ are the electric and magnetic vectors in the same medium.

In the general case, the propagation vector $\mathbf{k_0}$ (or the incident ray) is not necessarily in a direction perpendicular to the interface plane. This ray is said to be at oblique incidence. So the electrical field vector $\mathbf{E_0}$ is not always in the plane including the propagation vector $\mathbf{k_0}$ and the direction perpendicular to the interface, named the incidence plane (x, z) and represented on Figure 1.2. To see the behaviour of electrical field $\mathbf{E_0}$ at the air-glass interface, $\mathbf{E_0}$ has to be decomposed into two parts; one component in the incidence plane (x, z) which will be noted $\mathbf{E_P}$ and one component perpendicular to the incidence plane (x, y) noted $\mathbf{E_N}$. This decomposition is equivalent to separation of the incident wave into a TM wave (transverse magnetic) and a TE wave (transverse electric).

The reflected and transmitted fields can be completely known and if $\theta_0 \neq 0$ the coefficients of the field amplitude reflection are given by:

$$r_P = -\{\tan(\theta_0 - \theta)/\tan(\theta_0 + \theta)\}$$

$$r_N = \{\sin(\theta - \theta_0)/\sin(\theta + \theta_0)\} \tag{1.6}$$

where θ_0 and θ are the incident and refracted angles. Coefficients of the field amplitude transmission can be also defined.

For the particular case of normal incidence ($\theta_0 = 0$) the normal component is zero because the electric field is in the incident plane. In this case, the amplitude reflection coefficient is then reduced to the parallel component expressed by:

$$r = (n_0 - n)/(n_0 + n) \tag{1.7}$$

and for the optical power reflection coefficient R (also known as the *Fresnel reflection coefficient*), we find:

$$R = [(n_0 - n)/(n_0 + n)]^2 \tag{1.8}$$

and the optical power transmission coefficient T (or *Fresnel transmission coefficient*) is:

$$T = 4\,n_0\,n/(n_0 + n)^2 = 1 - R \tag{1.9}$$

1.1.2 Optical fibre

An optical fibre is a circular waveguide consisting of dielectrics of low optical loss. It comprises a central portion known as the core, surrounded by a cladding material (see Figure 1.4). The core has a refractive index n_1 higher than the refractive index n_2 of the cladding. It is used as a transmission medium for guided optical waves. The cladding is covered with protective plastic sheathing. The energy is carried partly inside the core and partly outside. The external field

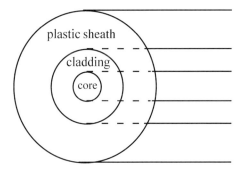

Figure 1.4. Schematic representation of an optical fibre.

decays rapidly to zero in the direction perpendicular to the propagation direction. Because the refractive index of the core is slightly higher than the cladding one, total reflection of most optical rays within the core occurs. Hence rays will be continuously reflected and travel a long distance within the fibre if they remain close to the axis and if they do not encounter sharp bends.

1.1.3 Step index multimode fibre

This consists of a homogeneous core of refractive index n_1 surrounded by a cladding of slightly lower refractive index n_2 (see Figure 1.5a). If r describes the radial distance from the core centre and a the core radius

$$n(r) = n_1 \quad a \leq r \leq 0$$
$$n(r) = n_2 \quad r > a$$

The relative index difference Δ is given by:

$$\Delta = (n_1^2 - n_2^2)/2n_1^2 \cong (n_1 - n_2)/n_1 \qquad (1.10)$$

Light is guided by total internal reflection. This is the oldest fibre type. Δ is of the order of 1 % or less for fibres based on fused silica. The core diameter (50–500 µm) is important. Several other fibre kinds are also available such as burried step profile fibres.

1.1.4 Graded index multimode fibre

Earlier it was noted that light rays can be deflected by variations in the refractive index of a medium as well by encountering an abrupt interface between two indices. In these fibres the light is not guided by total reflection but by a refractive index gradient (see Figure 1.5b). The

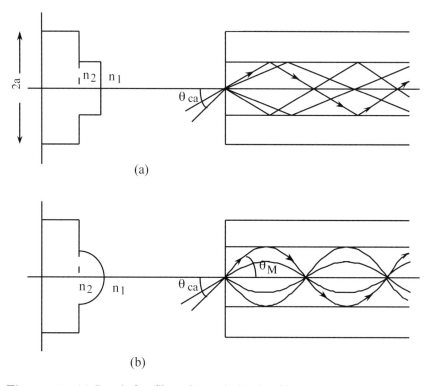

Figure 1.5. (a) Step-index fibre, (b) graded-index fibre.

optimum index profile is near-parabolic. If r describes the radial distance from the core centre, n_1 the refractive index at the core centre, a the core radius, the refractive index profile is:

$$n(r) = n_1 \left[1 - 2\Delta \left(\frac{r}{a}\right)^p\right]^{1/2} \quad r < a$$
$$n(r) = n_2 \quad r > a \quad (1.11)$$

$r^2 = x^2 + y^2$, x and y being the transverse coordinates. The parameter p is characteristic of the doped profile.

For a step-index profile

$$p = \infty \quad \text{for } r < a$$
$$p = 0 \quad \text{for } r > a$$

If $p = 2$ the index profile is parabolic.

1.1.5 Numerical aperture

The numerical aperture, (NA) is a parameter defining how much light can be collected by an optical system, which may be an optical fibre or a microscope objective. It is the product of the refractive index of the incident medium and the sine of the critical angle.

$$\text{NA} = n_0 \sin(\theta_{ca}) \quad (1.12)$$

In most cases, the light is incident from air and $n_0 = 1$.

The numerical aperture of a step-index fibre is given by the critical angle θ_{ca}, which can be used to find the size of the light cone that will be accepted by an optical fibre. In Figure 1.6 a ray is drawn incident on the core-cladding interface at angle α complementary to angle θ_c where $\theta_c + \alpha = \pi/2$. Then, if α corresponds to the critical angle θ_{ca}, we have:

$$\text{NA} = \sin\theta_{ca} = (n_1^2 - n_2^2)^{1/2} \quad (1.13)$$

When Δ is very small (weakly-guiding approximation), the NA is given approximately by:

$$\text{NA} \simeq (2n_1^2\Delta)^{1/2} = n_1(2\Delta)^{1/2} \quad (1.14)$$

In the same manner as for the case of step-index fibre a local numerical aperture can be defined for the graded-index fibre as follows:

$$\text{NA}(r) = n(r)\sin\theta_c(r) = [n^2(r) - n_2^2]^{1/2} \quad (1.15)$$

where $\sin\theta_c(r)$ is the maximum value for the ray to be guided.

1.1.6 Time delay

Because the refractive index n depends on the wavelength λ, pulse spread occurs in the optical fibre. This is partly caused by material dispersion. The material dispersion is given by:

$$M = (\lambda/c)(\mathrm{d}^2 n/\mathrm{d}\lambda^2) \quad ps.mm.km \quad (1.16)$$

A pulse propagating inside the optical fibre is thus lengthened if its spectral bandwidth is large. Another cause of pulse spread is the guiding effect. Consider (Figure 1.6) two rays in a step-index fibre. One, the axial ray, travels along the axis of the fibre; the other, the marginal ray, travels along a path near the critical angle for the core–cladding interface and is the highest-angle ray that can be propagated by the fibre. At the point where the marginal ray hits the interface, the ray

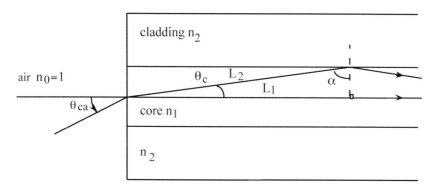

Figure 1.6. Optical path in a step-index fibre.

has travelled a distance L_2, while the axial ray has travelled a distance L_1. From the geometry, it can be seen that:

$$\sin\alpha = n_2/n_1 = L_1/L_2$$

The length L_2 is greater than L_1 by a factor n_1/n_2 in the case shown in Figure 1.6. For any length of fibre L, the additional distance travelled by a marginal ray is

$$\delta L = (n_1 - n_2)L/n_2 \cong L(n_1/n_2)\Delta$$

The additional time it takes light to travel along this marginal ray is:

$$\delta t = \delta L/v = L n_1 \Delta /c \qquad (1.17)$$

Therefore, a pulse with a length representing one bit of information will be lengthened to $t + \delta t$. This time difference between axial and marginal rays will cause the pulse to smear, and thereby limits the number of pulses per second that can be sent through a fibre and be distinguishable at the far end. In such a case the system may be limited not by how fast the source can be turned on and off or by the response speed of the detector, but by the differential time delay of the fibre. This smearing of pulses can be minimized or avoided by the use of graded-index or single-mode fibres.

1.2 PROPAGATION IN OPTICAL FIBRES

1.2.1 Multimode fibres

In this paragraph we recall the results for the modes of an ideal fibre in homogeneous, isotropic and linear media. Theoretical treatments can be found in Snyder and Love 1983, Jeunhomme 1988, Vassallo 1991. The model of light rays reflected at the boundary between core and cladding cannot entirely explain light propagation into fibres. The number of rays is limited; they do not form a continuous spectrum; and a speckle pattern appears at the end of a multimode fibre when a coherent source is used. That is why modes are introduced to describe propagation in fibres. A mode is a three-dimensional field configuration (**E**, **H**, **k**). It is characterized by a single propagation constant β connected to phase speed. It represents one of the possible solutions of Maxwell equations for the geometry and refractive index profile. In a fibre only a finite number N of modes exists. The Maxwell equations give the relations between each electromagnetic component (**E**, **H**) of the field.

$$\text{curl } \mathbf{E} = -\partial \mathbf{B}/\partial t \quad \text{curl } \mathbf{H} = -\partial \mathbf{D}/\partial t + \sigma \mathbf{E} \tag{1.18}$$

with $\sigma = 0$ for dielectrics.

$$\mathbf{D} = \bar{\varepsilon}\mathbf{E} \quad \mathbf{B} = \bar{\mu}\mathbf{H}$$

$\bar{\varepsilon}$ is the electric permittivity tensor and $\bar{\mu}$ the magnetic permeability tensor of each medium. The Fourier transform (*FT*) of each equation is taken with respect to t and the electric field is assumed to be sinusoidal in time so it can be written in complex notation in a cylindrical coordinate system as:

$$\mathbf{E}(r,\varphi,z,\omega) = \mathbf{E}(r,\varphi)\,e^{i(\omega t - \beta z)} \tag{1.19}$$

with

$$\mathbf{E}(r,\varphi,z,\omega) = FT_\omega[\mathbf{E}(r,\varphi,z,t)]$$

ω is the optical pulsation, β the propagation constant of the wave propagated along the *z*-axis, and r, φ are the cylindrical coordinates (r radius, φ azimuth) at the point M of the field.

In the case of homogeneous media and without non-linear effects (i.e., low intensity beam), $\bar{\varepsilon}$ is a constant ($=\varepsilon$) and under the hypothesis $\bar{\mu}=\mu=$ constant, the Maxwell equations become:

$$\text{curl } \mathbf{E} = -i\omega\mu\mathbf{H} \qquad \mathbf{D} = \varepsilon\mathbf{E}$$
$$\text{curl } \mathbf{H} = +i\omega\varepsilon\mathbf{E} \qquad \mathbf{B} = \mu\mathbf{H}$$

The conservation of electrical charges is expressed by div $\mathbf{B}=0$ and div $\mathbf{D}=\rho$ where the electrical charge density ρ is zero for the chosen materials.

Combining all these equations one obtains the vectorial form:

$$\nabla(\nabla\mathbf{E}) - \nabla^2\mathbf{E} = k^2\varepsilon_m\mathbf{E} \qquad (1.20)$$

where $\varepsilon_m = \varepsilon/\varepsilon_0 = n_m^2$, ∇ is the nabla operator, ∇^2 is the vectorial Laplacian. ε_m is the reduced permittivity and $k = 2\pi/\lambda$ is the free space propagation constant of light of wavelength λ.

If we suppose that $\Delta < 1\%$ we can use the scalar wave approximation. The error imposed on all mode characteristics remains below 0.1%. If E_z is the electric field component along the O_z axis we have:

$$\frac{\partial^2 E_z}{\partial r^2} + \frac{1}{r}\frac{\partial E_z}{\partial r} + \frac{1}{r^2}\frac{\partial^2 E_z}{\partial \varphi^2} + \left(k^2 n^2(r) - \beta^2\right) E_z = 0 \qquad (1.21)$$

We examine E_z solutions such as:

$$E_z = F(r)\cos v\phi$$

and follow the same procedure for H_z.

The scalar wave propagation equation can only be solved analytically for a few index profiles $n(r)$ such as step-index or infinite parabolic profile. In the other case of guided modes, only numerical solutions exist.

For the case of step-index profiles ($n(r)=n_1$ for the core $n(r)=n_2$ for the cladding), the solutions in the core (and in the cladding) are Bessel J_v (and Mac-Donald K_v) functions. The continuity of the fields imposes at the core-cladding interface the following dispersion equation:

$$\left(\frac{J_v'(u)}{uJ_v(u)} + \frac{K_v'(w)}{wK_v(w)}\right) \cdot \left(\frac{J_v'(u)}{uJ_v(u)} + (1-2\Delta)\frac{K_v'(w)}{wK_v(w)}\right) = \left(\frac{\beta v}{kn_1}\right)^2 \left(\frac{V}{uw}\right)^4$$
$$(1.22)$$

Optical fibre and propagation

where u^2 and w^2 are related to the propagation constant β of each mode by the relationships:

$$u^2 = a^2(k^2 n_1^2 - \beta^2) \tag{1.23}$$

$$w^2 = a^2(\beta^2 - k^2 n_2^2)$$

The constant term V introduced in (1.22) is called the *normalized frequency* and is defined by:

$$V^2 = u^2 + w^2 = a^2 k^2 (n_1^2 - n_2^2) = a^2 k^2 \, n_1^2 \, 2\Delta \tag{1.24}$$

The symbol $'$ designs the derivative of the Bessel functions with respect to r.

By analysing the equation (1.22) we can see that four types of modes can exist. Two kinds of modes exist if $v = 0$ and two if $v \neq 0$.

If $v = 0$, when the first term of equation (1.22) is zero the modes are $TE_{0\mu}$ transverse electric (the E_z longitudinal component is zero). When the second term of the same equation is set to zero, the modes are $TM_{0\mu}$ transverse magnetic ($H_z = 0$) with always $v = 0$. The Greek letter μ corresponds to the μth root of the dispersion equation (first or second term if *TE* or *TM*).

For the case $v \neq 0$, the components E_z and H_z are not equal to zero and the modes are the hybrids $HE_{v\mu}$ or $EH_{v\mu}$. (These hybrid modes are called $HE_{v\mu}$ if the E_z component is higher than H_z and $EH_{v\mu}$ if E_z is lower than H_z). A particular case is HE_{11} which corresponds to the first mode appearing during the propagation. If there is only the HE_{11} mode, the fibre is said to be single-mode, sometimes called 'monomode'.

An approximation can be made by combining $HE_{v-1\,\mu}$ and $EH_{v+1\,\mu}$ modes which have nearly identical propagation constants if $\Delta \ll 1$ (weakly guiding fibre) (Gloge 1971). It induces linearly polarized modes $LP_{\mu v}$ with only one component in Cartesian coordinates for the cross-section. $LP_{\mu v}$ are transverse electromagnetic waves. Both electrical and magnetical fields are perpendicular to the direction of propagation. Each of these $HE_{v-1\,\mu}$ and $EH_{v+1\,\mu}$ modes constituting $LP_{v\mu}$ are nearly degenerate. $LP_{v\mu}$ can be considered as modes having each two independent linear components if their azimuthal dependence is upon $\sin v\varphi$ or $\cos v\varphi$. Each mode is polarized vertically or horizontally (see Figure 1.7).

The normalized frequency V is important for characterization of propagation in fibres. It permits calculation of the number N of guided modes for a fibre with a large number of modes. We have:

Propagation in optical fibres

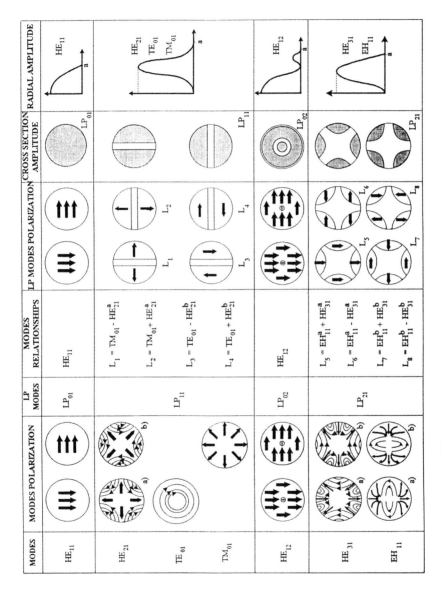

Figure 1.7. Scheme of polarized $LP_{\nu\mu}$ modes.

for step-index fibres

$$\frac{V^2}{2} - 1 < N < \frac{V^2}{2}$$

and

$$\frac{V^2}{4} - 1 < N < \frac{V^2}{4}$$

for graded-index fibres with parabolic profile. N includes both possible polarizations of the mode, $HE_{\nu\mu}$ (or $EH_{\nu\mu}$).

For example, in a step-index fibre with a normalized frequency $V = 3$ at a given wavelength λ, $3,5 < N < 4,5$. So there are 4 modes propagated by the fibre: HE_{11}, TM_{01}, TE_{01}, HE_{21}.

The cut-off normalized frequency of HE_{11} is 0. The HE_{11} mode is always propagated into the fibre. Indeed, the next modes appearing in the fibre are the TE_{01} and TM_{01} modes. These modes TE_{01} and TM_{01} correspond to the first root solution of $J_0(x) = 0$.

To eliminate all but the HE_{11} mode, the normalized frequency should be less than the one corresponding to the cut-off of the second mode TE_{01} or TM_{01}. This means $V < 2.404$ with $J_0(2.404) = 0$, first root of Bessel function. This assumption could be realised when the fibre core radius a or the index difference Δ are small enough. If $V \geq 2.404$, then several modes propagate.

Another useful quantity is often used to characterize each propagating mode, namely the effective index n_e connected to the mode propagating constant β by:

$$n_e = \beta/k = \beta\lambda_0/2\pi \qquad (1.25)$$

1.2.2 Single-mode fibre

Then multipath or multimode transmission can be avoided almost completely by choosing a core diameter so small that only one axial ray or one mode can propagate. These fibres are most useful where a large frequency bandwidth is needed. In order to operate in single-mode, the refractive profile index is generally step-index and the core radius a of the fibre has to be below a value a_c:

$$a_c = (1.202/\pi)(\lambda/2n_1) \qquad (1.26)$$

which corresponds to $V = 2.404$.

Spot size is used to quantify the distribution of power in the fundamental mode of single-mode fibres. There is a number of definitions of spot size. Peterman (1977) gives the definition of spot size ω_0 commonly adopted in terms of the normalized second momentum of the field, based on weak guidance.

The modal distribution can be satisfactorily approximated by a Gaussian function $\psi_g = \psi_{go} \exp[-(r/\omega_o)^2]$ with ω_o permitting to maximize the overlap integral χ

$$\chi = \left| \int_0^\infty E(r)\psi_g r\, dr \right|^2 \bigg/ \left| \int_0^\infty |\psi_g|^2 r\, dr \int_0^\infty |E(r)|^2 r\, dr \right|$$

$E(r)$ is the real expression of the field; ω_o is so defined by $d\chi/d\omega_o = 0$. Marcuse (1977) proposes an analytical expression which gives a good approximation of ω_o versus λ or V corresponding to amplitude field width at e^{-1}:

$$\omega_o/a = 0.65 + 1.619\, V^{-3/2} + 2.879\, V^{-6} \tag{1.27}$$

with

$$V = ak\,\mathrm{NA} = (2\pi/\lambda)\, a\, n_1 (2\Delta)^{1/2}$$

1.3 CHARACTERISTICS AND MEASUREMENTS

1.3.1 Numerical aperture

In step-index fibre, numerical aperture is defined by $\mathrm{NA} = n_1 (2\Delta)^{1/2} = (n_1^2 - n_2^2)^{1/2}$ (see equation 1.13) as shown in § 1.1.5. It is related to the fibre maximum acceptance angle which is the maximum angle θ_{ca} for rays propagating in air and injected into the fibre (equation 1.12).

For the case of graded-index fibres, a local numerical aperture is defined by:

$$\mathrm{NA} = n(r) \sin \theta_c(r)$$

For the central point ($r = 0$) the same relation as for step-index fibres stands for numerical aperture. A typical method to obtain the numerical aperture measurement is the analysis of the far field pattern.

Two different ways of launching light into the fibre for measurement of numerical aperture are shown in Figure 1.8. In (a) a beam of light is focused into the fibre by a microscope objective with an NA of not less than the estimated NA of the fibre. At the other end of the fibre a detector centred on the end-face moves through an angle θ, measuring the light intensity. In (b) the detector remains aligned with the fibre while a plane wave wider than the fibre core moves through the angle θ.

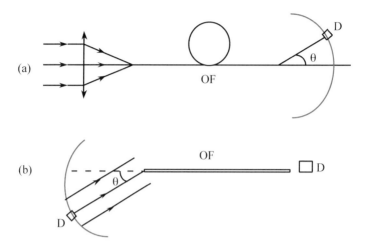

Figure 1.8. Measurements of numerical aperture: O microscope objective, D detector, OF optical fibre (a) with focused light, (b) with plane wave.

Generally NA is calculated by taking the angle at which the accepted power has fallen down to 95 % of the maximum accepted power at $\theta = 0$. Typical fibre ($\Delta \cong 0.01$, $n_1 = 1.46$ at $\lambda = 630$ nm) gives the numerical value NA $= 0.2$ which corresponds to $\theta = 11.5°$. NA range values are from about 0.1 for a single mode fibre to 0.2–0.3 for multimode communication fibres and up to 0.5 for large core fibres.

1.3.2 Diameter control and measurement

In practice, the mode propagation in an optical fibre is affected by core diameter variations. One method for measurement and controlling of the core diameter is based on refracted rays interference. A laser beam is launched perpendicular to the fibre propagation axis. The optical path difference between the reflected ray on the external surface and the refracted ray at the air–cladding interface creates interferences. The fringe spacing is proportional to the fibre diameter (Watkins 1974). Study of the fringe system gives precise information on the fibre geometry (Gagnaire *et al.* 1981) (see Figure 1.9).

1.3.3 Attenuation measurement

Attenuation proceeds from diffusion and absorption losses. The power P after a fibre length L is related to the injected power P_i by:

$$P = P_i \exp[-\alpha_1 L] \tag{1.28}$$

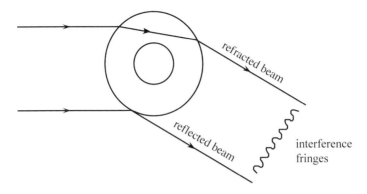

Figure 1.9. Experiment for diameter measurement.

The α_1 attenuation coefficient converted into α coefficient is expressed in decibels per kilometre:

$$\alpha(\text{dB/km}) = (10/L)\log_{10}(P_i/P) = 4.34294\,\alpha_1(\text{in km}^{-1}) \quad (1.29)$$

Good quality glasses have an attenuation coefficient of about 1000 dBkm^{-1}. For optical fibres, losses have to be much reduced. By suppression of impurities 0.01 dBkm^{-1} has been obtained.

Rayleigh scattering in dielectric media is due to micro irregularities acting like so many diffraction centres. This phenomenon follows a $1/\lambda^4$ law. That is why working at longer, infrared wavelengths is very useful. Absorption can appear when core material contains transition metals in which the electron energy levels are not completely full, so their presence has to be avoided. This is technologically easier than suppressing numerous OH$^-$ ions. It is achieved by drying the fibre in an atmosphere of chlorine. Losses such as those introduced by bending or micro-bending need to be avoided.

1.3.3.1 Cut-fibre method

The power $P(z,\lambda)$ for different values of length z at a given wavelength is measured and plotted on a semi-log scale. The slope of the curve gives the value of $\alpha/10$ (α: attenuation coefficient in dB). More commonly two measurements are taken, one with the total fibre length, the other with a measured length of several meters of the same fibre. This method is not currently used because the fibre has to be cut.

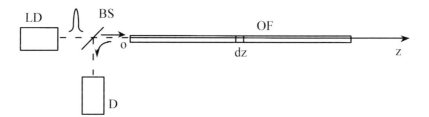

Figure 1.10. Retrodiffusion method: BS Beam Splitter, D Detector, OF Optical Fibre, LD Laser Diode.

1.3.3.2 Retro-diffusion method

A light pulse is injected into an optical fibre (see Figure 1.10). A proportion of diffused light in the optical fibre creates feedback and the echo continuation forms an envelope sent back to the detector and analysed at the fibre entry face. The back-scattered signal P(z), a function of the distance for a symmetric fibre of constant attenuation α, is written as:

$$P(z) = 0.5 P_o \tau S \alpha_d V_{gr} \exp(-2\alpha z) = A \exp(-2\alpha z)$$

where P_o is the injected power, τ the light pulse width, α_d the diffusion coefficient, S the total solid angle, V_{gr} the group velocity. The signal $P_c(z)$ reflected by a break of reflecting power R is:

$$P_c(z) = R P_o \exp\left(-\int_0^z (\alpha'(l) + \alpha''(l)) \mathrm{d}l\right)$$

where α' and α'' are the attenuations for forward and backward paths. Then the signal $P_c(z)$ can be compared with the retrodiffusion signal just forward of the break.

Other methods can be used. The lateral scattering is measured by the power lost along the fibre. The reflection pulse is realized by setting a mirror at the end of the fibre. The last method we will mention is the differential modal attenuation. Plane waves with different angles θ or a focused wave are injected in a step or a graded index fibre. Either or both θ and r can be varied. Each mode μ is given by the relation:

$$\mu = 2\pi[(r/a)^p + (\sin\theta / \sin\theta_{ca})^2]^{(p+2)/2p}$$

with p the profile parameter, a the core radius, θ_{ca} the acceptance angle, r the radius coordinate and θ the angle coordinate. The enlarged image of the fibre end is formed on a diaphragm in front of a detector. Several fibre parts are selected. The detector measures the emitting power for varying values of θ.

1.4 BIREFRINGENCE

An important activity area in optical fibres is the development of polarization-maintaining fibres. These are single-mode fibres able to preserve a stable state of linear polarization. This ability is derived from the intrinsic birefringence which is designed to be greater than the extrinsic environmentally induced birefringence. Intrinsic birefringence can be achieved by core asymmetry or an asymmetric stress distribution in the core, using either an elliptical core or a mechanical stress profile to induce linear birefringence.

We can thus distinguish two types of birefringence. Intrinsic birefringence appears in a fibre because of built in anisotropies: shape or stress. An induced birefringence appears when a fibre is bent, twisted or otherwise submitted to forces or to external fields.

1.4.1 Shape birefringence

Consider a single mode fibre with an elliptical core whose major axis is $2a$ and minor axis $2b$ (see Figure 1.11a). In such a fibre the x-polarized and y-polarized LP_{01} modes will have slightly different propagation constants β_x and β_y. We define birefringence:

$$B = \delta\beta/k \qquad (1.30)$$

where

$$\delta\beta = \beta_y - \beta_x$$

As soon as the two LP_{01} modes have different propagation constants, they have different propagation group delay times. A very strong birefringence requires both a large major–minor axis ratio and a high index difference. The beat length L_b is the fibre length such as:

$$\delta\beta L_b = 2\pi \qquad (1.31)$$

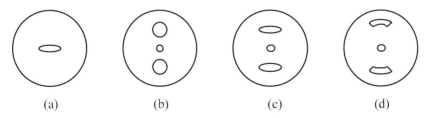

(a)   (b)   (c)   (d)

Figure 1.11. Birefringent fibres: (a) with elliptical core, (b) (c) with stress regions: PANDA, (d) with stress region: Bow-tie.

20 *Optical fibre and propagation*

The extinction ratio η is the fraction of the total cross-power coupled into the x-polarization component from the y-polarization component for y-polarization eigenmode excitation:

$$\eta = I_x/(I_x + I_y) \tag{1.32}$$

High birefringence optical fibres are also obtained by preform deformation (Stolen et al. 1984) and have a rectangular geometric cross section.

1.4.2 Stress birefringence

In birefringent fibres we have two principal polarization axes. These two axes may be due to stress application (see Figure 1.11b, c, d). In a "panda fibre" as in "Bow tie" fibre, the stress is due to doped stress areas near the core.

1.4.3 Stokes parameters and Poincaré sphere

The problem to measure more general states of polarization is soluble using readily obtainable techniques. They involve essentially the measurement of the so-called Stokes parameters of the incident polarized wave. In order to determine these parameters the incident wave is crossing four devices which measure various polarization properties of the wave:

- polarization-insensitive half power attenuator
- linear polarizer oriented horizontally
- linear polarizer oriented at 45° from the horizontal
- right-handed circular polarizer.

The intensities of the transmitted light through these four devices are I_0, I_1, I_2, I_3 and the Stokes parameters are:

$$\begin{aligned} S_0 &= 2I_0 \\ S_1 &= 2I_1 - 2I_0 \\ S_2 &= 2I_2 - 2I_0 \\ S_3 &= 2I_3 - 2I_0 \end{aligned} \tag{1.33}$$

S_1 measures the amount of linear horizontal polarization, S_2 the amount of linear polarization at 45° from the horizontal, S_3 the circularity of the incident beam. They characterize the state of polarization (SOP) and are normalized such that $S_0 = 1$. Linearly horizontally polarized light corresponds to $S_1 = 1$, $S_2 = S_3 = 0$; vertical polarized light to $S_1 = -1$, $S_2 = S_3 = 0$. If $S_3 = 1$, $S_1 = S_2 = 0$ the light is right circularly polarized and if $S_3 = -1$, $S_1 = S_2 = 0$ the light is left circularly polarized.

If we represent elliptical polarization (see Figure 1.12a) in a reference system $0x$, $0y$, it may be characterized by the aspect ratio $\tan v = b/a$, where b and a are the lengths of the two axes, and by angle ω between the axis of the ellipse and the x-axis. The Stokes parameters are

$$S_1 = I \cos 2v \cos 2\omega$$
$$S_2 = I \cos 2v \sin 2\omega \qquad (1.34)$$
$$S_3 = I \sin 2v$$

A useful representation of the Stokes parameters is the Poincaré sphere. This geometrical construction permits a simple calculation of the evolution of the state of polarization (SOP) of an arbitrary input polarization through an arbitrary birefringent medium (see Figure 1.12b).

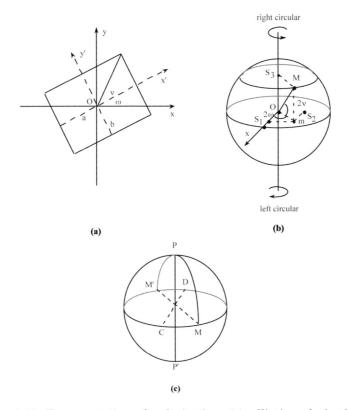

Figure 1.12. Representation of polarization: (a) elliptic polarization, (b) Poincaré sphere representation of a point M, (c) transformation by a quarter and by a half wave plate on the Poincaré sphere.

The equator on the sphere represents linear polarization states ($S_3 = 0$, $v = 0$). The north and south poles on the sphere represent right and left circularly polarized light respectively. An arbitrary origin may be chosen as the x-axis (horizontal polarization). Linear y-polarization (vertical polarization) is represented by the point y on the sphere diametrically opposite. Lines of constant latitude on the sphere describe SOPs of constant ellipticity e given by $e = \tan v$. The Stokes parameters S_1, S_2, S_3 and the angles v and ω are reported on the Poincaré sphere (see Figure 1.12b).

An arbitrary birefringent section may be characterized in terms of the eigenmodes through that section and of the phase difference for propagation of these eigenmodes. These eigenmodes are the SOP$_s$ which pass through the birefringent section unaltered and are elliptical. They are represented by diametrically opposite points on the Poincaré sphere. The phase difference between the two eigenmodes being θ, the position of the input polarization on the sphere, after the sphere has been rotated through angle θ about the diameter joining the eigenmodes, represents the output polarization. Conversely, any rotation around an arbitrary axis on the sphere may be resolved into polar and equatorial rotations, and any path may be resolved into circularly birefringent components with eigenmodes on the poles and a linearly birefringent component with eigenmodes on the equator.

A quarter-wave plate would have the points C and D as eigenmodes. If the input polarization to a quarter-wave plate is linear and at 45° to the principal axes – point M on the sphere – then rotating the sphere by 90° puts point M on to one or other of the poles P or P', representing circular polarization. Putting circular polarization into a quarter-wave plate rotates the pole on to the equator at an angle exactly bisecting the principal axes, so that linear polarization emerges.

A half-wave plate–with the same eigenmodes, but a 180° phase difference–will always produce linear output polarization from linear input, taking the input to an output that is the reflection of the input in a plane defined by the poles and the diameter (point M' on Figure 1.12c).

1.5 FABRICATION AND MATERIALS

1.5.1 Fabrication

There is a variety of methods of creating controlled index gradients. Some involve introducing impurities into thin layers of glass as they are laid down on a substrate. This is not a continuous process since the refractive index within each layer is nearly constant. The resulting variation of refractive index in a fibre resembles a set of concentric rings rather than a smooth change in the index.

Silica fibres are made by drawing a preform which has a structure similar to that of the required fibre. There are various methods of preparing performs, namely MCVD (modified chemical vapour deposition), OVD (outside vapour deposition), VAD (vapour axial deposition). These methods are the most popular processes. Details on materials and fabrication can be found in Izawa and Sudo (1987). In this chapter we will discuss only the MCVD method. A silica (SiO_2) film is deposited on the inside surface of a fused silica tube (see Figure 1.13).

This tube is put on a glass working lathe and is rotated. Raw halide material vapors carried by oxygen gas are introduced into the silica tube. A driven carriage which carries oxy-hydrogen torches at a constant speed is added to the lathe. The flame heats the tube from the outside to about 1600 °C. It is moved repeatedly along the tube, and during each traverse the halide vapours are oxidized to fine glass particles and deposited on the inner surface of the tube. The heated zone can be kept at a constant temperature by controlling the flow rates of oxygen and hydrogen. The raw materials used for high silica glass are silicon, germanium and boron chlorides and phosphorus chlorate. To reduce the refractive index to lower than that of silica, dopants such as $SiCl_4$ or BCl_3 are used; for higher values, $GeCl_4$ or $POCl_3$ are used. The flow rate is controlled by a gas supply system and the mixing ratio of halide vapour is controlled to a programmed value. By changing $GeCl_4$ concentration in the vapours for each traverse, the refractive index of each glass layer can be controlled. The metal halides (except BCl_3) remain in the liquid state at less than 50 °C giving porous material. Dehydratation and consolidation follow the process. The reacted gas is exhausted from the silica tube. After sufficient film thickness is obtained, the tube is collapsed so that the deposited layers form a high index core and the tube provides the low index cladding.

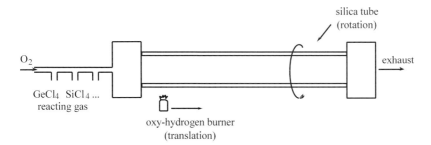

Figure 1.13. Drawing apparatus.

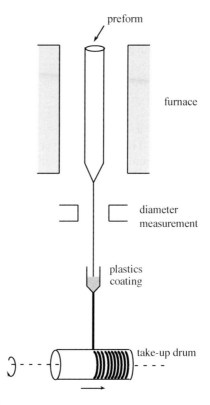

Figure 1.14. Drawing apparatus.

This preform is drawn into a fibre in a furnace. The drawing equipment includes a preform carrier, electric furnace, outer diameter measurement instrument, plastics coating applicator and take-up drum (see Figure 1.14).

1.5.2 Materials

The three major material systems used in the manufacture of fibres are silica, others glass types and plastics.

The silica system is a mixture of silicon dioxide, SiO_2, and other metal oxides, to establish a difference in refractive index between the core and cladding. Various dopants have been used to increase the refractive index of silica including titanium, aluminium, germanium, and phosphorus oxides. Chalcogenide glasses have a longer cut-off wavelength than oxide and fluoride glasses. They are solid solutions of metal sulphides, selenides and tellunides of arsenic, germanium and antimony. They have a stable vitreous state and a wide range of transmission wavelengths.

Fluoride glasses are made from beryllium, zirconium, mercury, aluminium and barium fluorides. They are promising materials for use in the infrared wavelength range because they have very low losses in this region, up to 2 μm, whereas losses in oxide glasses is comparatively high.

The advantages of plastics fibres are low price and ease of use due to large core diameters. They have a much higher attenuation coefficient than glass fibres. Their length can reach 50 m. The numerical aperture is around 0.5 and the attenuation from 10 to 1000 dBkm^{-1}. Typical plastics used for optical fibres are poly-methyl methacrylate and polystyrene. The transmission loss of plastics results mainly from vibrational absorption owing to carbon–hydrogen bonds and Rayleigh scattering.

1.5.3 Rare-earth doped fibres

In addition to silicate and fluoride glasses, phosphate, fluorophosphate fluorozirconate, fluoroberyllate and ZBLAN (zirconium, barium, lanthanium, sodium) glasses are also used as host materials for rare-earth doped fibre. These particular fibres, when used as laser fibres and amplifiers (Ainslie *et al.* 1990, Miniscalco 1991, Cognolato and Gnozzo 1993) have their core doped (for example with lanthanide ions with incomplete inner 4f levels). Neodynium (Nd^{3+}) and Erbium (Er^{3+}) are the most common rare earth used as dopants. Other ions like Ce^{3+}, Pr^{3+}, Tb^{3+}, Dy^{3+}, Ho^{3+}, Tm^{3+} and Yb^{3+} can be used in host materials like heavy metal fluoride glass fibres (ZBLAN, ZBLANPb, ZBLAN P). All these fibres can be fabricated by rare-earth vapour phase deposition at high pressure, or by impregnation and diffusion of rare-earth doped solutions followed by drying. Generally, the concentration of rare-earth materials is less than those of the classical dopants (300–1000 ppm) for the highest gain.

1.6 LIGHT SOURCES

With optical fibre, suitable light sources are lasers, LEDs and sometimes filament lamps. The model of a Gaussian beam is used to characterize the sources radiation geometry specifically for lasers. For LEDs the Lambertian model of sources is used.

1.6.1 Laser diodes

The Gaussian beam is the simplest beam that complies with Maxwell's equations. It has linear polarization with its electric field orientated along the y-direction, perpendicular to the direction of propagation z. The magnetic field is orientated along the x-direction. It has a finite width that smoothly transforms along the z-axis into a narrow one of fixed numerical aperture (see Figure 1.15).

The electrical field is expressed as:

$$E(r,z) = E_0(z) \exp[-r^2/\omega^2(z)] \qquad (1.35)$$

z is the distance from the beam waist, along the propagation direction, $E_o(z)$ the electric field, r the radial distance from the z-axis, $\omega(z)$ the beam radius where the power density is down to $1/e^2$ of maximum (the electrical field is down to $1/e$ of maximum).

At z, the beam radius $\omega(z)$ is related to the beam waist radius ω_o:

$$\begin{aligned}\omega^2(z) &= \omega_0^2(z)[\,1 + (\lambda z/\pi \omega_0^2)^2] &&\text{for small } z \\ \omega(z) &= \lambda z/\pi\, \omega_0 &&\text{for large } z\end{aligned} \qquad (1.36)$$

ω_o is the radius of the beam waist. At the beam waist the wave front is planar. At larger distances it becomes spherical, and the beam can be described by a numerical aperture:

$$\mathrm{NA} = 0.18\, \lambda/\omega_0 \qquad (1.37)$$

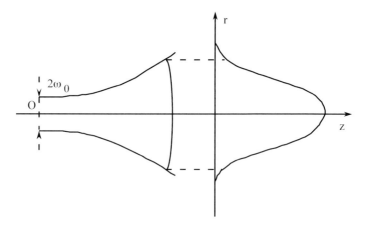

Figure 1.15. Propagation of a Gaussian beam: ω_0 is the radius at the beam waist, z the direction of propagation.

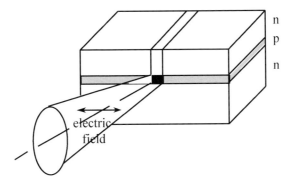

Figure 1.16. Representation of the laser diode with the radiation field.

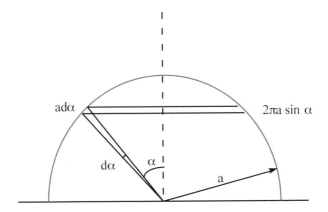

Figure 1.17. Measurement of the power density.

For this Gaussian beam, the total power P and the on-axis irradiance $M(z)$ are found by integrating the irradiance on a plane located at z from the origin. We have:

$$P = (\pi/2)M(z)\omega^2(z)$$

Laser diodes are sources used in telecommunications and sensors. In typical laser diodes, the active zone is 4–12 μm wide, 0.1–0.3 μm high and 300 μm long (average) value. The facets behave as semi-transparent mirrors (30% reflectance) at the ends of the resonator cavity. Emission is through both facets. Lasers generate partially polarized light. The orientation of the electrical field is parallel to the junction. The radiation pattern in the far field is elliptical with a full width at half maximum between 20° and 40°. The ellipticity stems from different degrees of diffraction at parallel and perpendicular junction cavity boundaries. The near field of a laser diode is elliptical too (see Figure 1.16).

1.6.2 LED sources

In the case of a Lambertian source defined by an uniformly diffusion surface and corresponding to LED's emission model, let us consider a source with a small diameter and a detector with a small surface that is assumed to move at a constant distance, the centre being at the source location. The viewing angle is α. On the axis (i.e., perpendicular to the source) we measure a power density M_0. In the field, when leaving the axis, the measurements follow a cosine relationship (see Figure 1.17). We have:

$$\text{Irradiance}(W\ m^{-2}) \qquad M(\alpha) = M_0 \cos \alpha$$

$$\text{Radiance}(W\ sr^{-1}\ m^{-2}) \qquad L(\alpha) = L_0 \cos \alpha$$

$$\text{Intensity}(W\ m^{-2}) \qquad I(\alpha) = I_0 \cos \alpha$$

a being the source radius, the total power in all space is given by:

$$P = \int_0^{\pi/2} M(\alpha) 2\pi a \sin \alpha d\alpha$$

$$P = \pi a\, M_0 \quad \text{with} \quad M_0 = \pi a\, I_0$$

This relation can be used to estimate LED power. Most of the optical sources can be modelled by a raised cosine function (Goure and Massot 1982).

REFERENCES FOR CHAPTER 1

Adams MJ 1981 *An introduction to optical waveguides* (Wiley: New York)
Ainslie BJ, Craig-Ryan SP, Davey ST, Armitage JR, Atkins CG, Massicott JF and Wyatt R 1990 *IEE Proc. PtJ* **137** 205–8
Arnaud JA 1976 *Beam and fiber optics* (Academic Press: New York)
Cognolato L and Gnazzo A 1993 *Optical Materials* **2** 1–9
Culshaw B 1984 *Optical fibre sensing and signal processing* (Peter Peregrinus: London)
Dakin JP and Culshaw B 1988 *Opt. Fiber sensors I Principles and components* (Artech House: London)

Dakin JP and Culshaw B 1989 *Opt. Fiber sensors II Systems and applications* (Artech House: London)

Gagnaire H, Goure JP and Massot JN 1981 *Opt. and Quant. Elect.* **13** 55–64

Gloge D 1971 *Appl. Opt.* **10** 2252–8

Goure JP and Massot JN 1982 *Opt. and Quant. Elect.* **14** 445–51

Goure JP, Verrier I and Meunier JP 1989 *J. Phys. D: Appl. Phys* **22** 1791–805

Izawa T and Sudo S 1987 *Optical fibers: materials and fabrication* (KTK Scientific Publishers)

Jeunhomme LB 1988 *Single-mode fiber optics* (Marcel Dekker: New York)

Jones BE (ed) 1987 *Current advances in sensors* (Adam Hilger: Bristol)

Kapany NS and Burke JJ 1972 *Optical waveguides* (Academic Press: New York)

Marcuse D 1974 *Theory of dielectric optical waveguides* (Academic Press: New York)

Marcuse D 1977 *Bell Syst. Tech. J.* **56** 703

Miller CM, Mettler SC and White IA 1986 *Optical fiber splices and connectors. Theory and methods* (Marcel Dekker: New York)

Miller SE and Kaminow IP 1988 *Optical fiber telecommunication II* (Academic Press, New York)

Miniscalco WJ 1991 *J. Lightwave Technol.* **9** 234–50

Neumann EG 1988 *Single-mode fibers – fundamentals* chapter 12 (Springer Verlag)

Noda J and Yokohama I 1988 *Fiber devices for fiber sensors Technical Digest II* **468** (Washington DC OSA)

Peterman F 1977 *Opt. and Quantum Electron.* **9** 167–75

Senior JM 1992 *Optical fibre communications: principle and practice* (Prentice Hall – International series in opto-electronics second edition)

Snyder AW and Love JD 1983 *Optical Waveguide Theory* (Chapman and Hall: London)

Stolen RH and De Paula RP 1987 *Proc. IEEE* **75** 1498–511

Stolen RH, Pleibel W and Simpson JR 1984 *J. Lightwave technol.* **LT 2** 639–41

Vassallo C 1985 *Théorie des guides d'ondes électromagnetiques Tome 1, 2* (Eyrolles: Paris)

Vassallo C 1991 *Optical waveguide concepts Optical wave sciences and technology Vol. 1* (Elsevier: New York)

Watkins LS 1974 *J. Opt. Soc. Am.* **64** 767–72

EXERCISES

1.1 – The refractive index of a fibre core is $n_1 = 1.46$. What is the value of the optical power reflective constant and the relative transmitted power when the fibre is lighted in air ($n_0 = 1$).

1.2 – Consider an optical step index fibre with refractive indices $n_1 = 1.540$ and $n_2 = 1.530$.
Calculate:
    a – the relative index difference
    b – the numerical aperture
    c – the critical angle θ_{ca}
    d – the difference of length and of propagation time between a ray propagating along the axis and a ray at the critical angle for 1 km of optical fibre.

1.3 – Given $P(0) = 1$ mW signal at the input of an optical fibre:
    a – what is the transmitted light power at a distance $l = 10$ km, if the attenuation is 0.2 dBkm^{-1}?
    b – What is the maximum allowable distance to the detector with a minimum reliable level of 1 μW?

1.4 – A fibre has the following characteristics:
NA $= 0.1$
$a = 5.93$ μm
$\Delta = 2.5 \times 10^{-3}$
Calculate the value of the wavelength in order to have a single mode fibre and the values of core and cladding refractive indices.

1.5 – Establish the wave equation for the magnetic field **H**.

Chapter 2

COUPLING: MICROCOMPONENTS, TAPERS, SPLICES, CONNECTORS

The increasing use of optical fibres for light wave communications, local area networks, or sensors, requires techniques for splicing and coupling. The implementation of optical fibres depends to a large extent on the availability of low cost connectors and devices in order to couple a single-mode fibre to integrated optical components or to couple small sources (laser diode or LED) to optical fibres. Major problems are difficulties in establishing and maintaining a precise and critical alignment. This requires careful mechanical polishing of waveguide ends. A further problem is the difference between diameters of the two parts (for example between single-mode and multimode fibre) or between sections of two light spots (for example between a circular fibre and a rectangular waveguide).

Coupling between a semiconductor laser and a single-mode fibre, or between any two optical elements, is often a coupling between two Gaussian beams. A review of coupling devices for single-mode fibres was carried out by Khoe *et al.* (1984) and in single mode fibre optical components by Minowa *et al.* (1982).

In this Chapter we describe some microcomponents such as GRIN lenses, spherical microlenses, micro Fresnel lenses and tapers. Then we present the methods used in coupling from fibre to fibre (splices, microcomponents, connectors), from fibre to waveguide and from sources (lasers and LEDs) to fibre.

2.1 FIBRE ENDS

For splicing two fibre ends or for coupling light into a fibre, flat end faces are required. There are several methods of preparing fibre end faces. The first method is grinding and polishing. The fibre end is embedded in

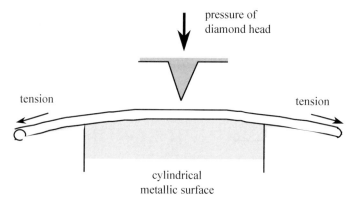

Figure 2.1. Principle of fibre tool.

epoxy resin or a metallic ferrule and polished using abrasives. Commercial equipment gives good durable quality, but this method is expensive and time consuming.

A second method is fibre cleaving (see Figure 2.1). The fibre is bent under tension and scribed with a small carbide, sapphire or diamond head. A wide range of cleaving tools is available. High quality ends perpendicular to the fibre axis can be obtained, with angles not departing from 90° by more than 1°. The end face quality can be checked with a microscope.

At the silica fibre end face in air the power reflection due to the Fresnel coefficient R (see equation 1.8) is around 3.5%. In many cases (coupling to laser source, interferometric sensors, backscattering measurements) this power loss cannot be tolerated. For these applications it is important to reduce the reflection factor. The end face can be coated by an anti-reflective multilayer dielectric or can be immersed in a liquid whose refractive index matches that of the fibre core (an *index-matching fluid*).

Another method of avoiding reflection losses is to polish the fibre obliquely. This method is used in fibre connectors (see § 2.3.4, Figure 2.20). Results of coupling efficiency have been presented for several end face angles by Clement et al. (1993).

In other applications, such as fibre resonators or in optical time domain reflectometer (O.T.D.R.) measurement technology, reflecting fibre ends are obligatory. A simple way of making an optical mirror is by plating the fibre end with an metal alloy having a low melting point. The obtained reflectance is 50–60% at $\lambda = 0.633$ μm (Ishikawa et al. 1985).

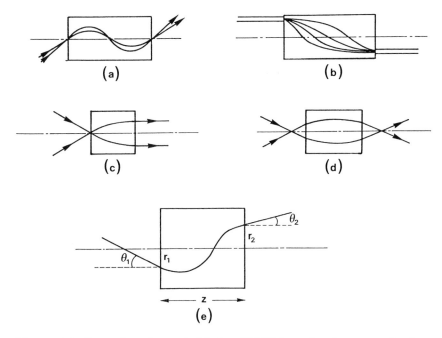

Figure 2.2. Operation of a graded-index (GRIN) lens showing ray paths for a object point: (a) on-axis (meridional rays), (b) off-axis (coupling between two fibres), (c) (d) used as relay lens, (e) input and output characteristics (r, θ) of a ray.

2.2 MICROCOMPONENTS

2.2.1 GRIN Lenses

Various methods have been proposed to reduce the coupling losses in optical fibre devices. Among the options, coupling using GRIN (for GRaded INdex) lenses offers great potential. These lenses are also called SELFOC^{TM} lenses. (SELFOC is a brand name of the Nippon Sheet Glass Co.)

In Chapter 1 we described graded index fibres. The index variation with radial distance in the core is given by the power law profile (Equation 1.11). A fan of rays injected into a graded-index fibre is converged to recross the axis at a common point in a similar way to rays from a small object reimaged by a lens (see Figure 2.2a). The distance it takes for a ray to travel one full sine path is called the *fibre pitch*. The pitch length is determined by the fractional index difference Δ. If a parabolic graded-index fibre is cut to a length of a quarter of the fibre pitch, it can serve as an extremely compact lens (see Figure 2.2c and d). It is called a GRIN lens and it is used in various fibre applications such as connecting. By positioning the output of a first fibre in front of the GRIN lens, light will be collimated, just as light diverging from the focal point of a lens is collimated (see Figure 2.2c). Because its properties are set by its length, this lens is referred to as a quarter-pitch or 0.25 pitch lens.

Equation (1.11) can be rewritten with $p = 2$ and $A = 2\Delta/a^2$ as:

$$n(r) = n_1[1 - Ar^2]^{1/2} \tag{2.1}$$

The gradient constant A and n_1 are functions of wavelength. When a meridional ray strikes the GRIN rod lens at position r_1 and angle θ_1 relative to the rod axis, position r_2 and angle θ_2 of the output ray can be given approximately as (see Figure 2.2e):

$$\begin{vmatrix} r_2 \\ \theta_2 \end{vmatrix} = M \begin{vmatrix} r_1 \\ \theta_1 \end{vmatrix} \tag{2.2}$$

where M is a ray matrix:

$$M = \begin{vmatrix} \cos(\sqrt{A}z) & \dfrac{1}{n_1\sqrt{A}}\sin\sqrt{A}z \\ -n_1\sqrt{A}\sin(\sqrt{A}z) & \cos(\sqrt{A}z) \end{vmatrix} \tag{2.3}$$

with z the length of the rod lens. The refractive index at the rod axis n_1 and the refractive index difference coefficient A are usually fixed and the rod lens length is treated as an independent parameter.

In equations (2.2) and (2.3), taking a full pitch length $Z = 2\pi/\sqrt{A}$ we have $r_2 = r_1$ and $\theta_2 = \theta_1$; taking a length $z = Z/2 = \pi/\sqrt{A}$ we obtain $r_2 = -r_1$ and $\theta_2 = -\theta_1$ (half pitch lens). A quarter pitch lens has a length $z = Z/4 = \pi/2\sqrt{A}$ and $r_2 = \theta_1/n_1\sqrt{A}$, $\theta_2 = -n_1\sqrt{A}r_1$.

An important factor in judging the transmittance quality of the branching components is the coupling efficiency. It may be reduced by aberrations, mismatch losses, misalignment losses and crosstalk. In fibre devices using GRIN rod lenses, the lenses form an image of an input fibre end face (or LD) on an output fibre end face. Aberrations broaden that image, reducing the efficiency.

Approximate expressions for aberration losses due to mismatch and misalignment for branching components, crosstalk due to reflection at the endface of rod lenses, as a function of the basic lens and fibres parameters, can be obtained using ray tracing calculations, geometrical optics, Gaussian beams and aberration theory results (Maekawa and Azuma 1987, Tomlinson 1980, Chen et al. 1992). GRIN

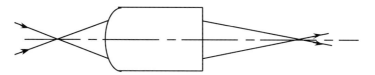

Figure 2.3. Plano-convex GRIN lens.

lenses are fabricated by the same methods used for graded index fibres or by ion exchange technology (Makita *et al.* 1988, Gomez-Reino 1992).

Increase of the NA and minimization of the wavefront aberration of the lens is necessary to obtain high efficiency. As well as conventional GRIN lenses, PC-GRIN lenses (Plano-Convex GRIN lenses) has been used (Makita 1988, Ishikawa *et al.* 1985) (see Figure 2.3). The convex surface increases the Numerical Aperture N.A and allows more power to be coupled into the lens. It decreases wavefront aberrations, resulting in a smaller focal spot on the fibre.

It should be noted that the imaging quality of the GRIN lenses is not as good as that of a good conventional lens, but coupling efficiency is equivalent or better, and leads to excellent solutions for photometric problems.

2.2.2 Microlenses

Another type of component used to obtain high coupling efficiency between LD and fibre or between fibre and fibre is the microlens. Several designs are available, and various coupling structures are described in the literature (see Figure 2.4).

Among these microlenses some are formed on fibre ends (Sakaguchi *et al.* 1981, Yamada *et al.* 1980, Eisenstein and Vitello 1982, Cohen and Schneider 1974, Tanigami *et al.* 1989, Kotsas *et al.* 1991, Eftimov and Hitchen 1993, Saitoh *et al.* 2000). Single mode fibres with tapered hemispherical or hemicylindrical have been also described (Mathyssek *et al.* 1985, Kuwahara *et al.* 1980, Khoe *et al.* 1983, Wang *et al.* 1991).

The use of spherical lenses has been proposed by Lambert and Goure (1985), Goure *et al.* (1984), GRIN rod lenses by Sugie and Saruwatari (1983), optical fibre taper by Presby *et al.* (1987) and non-imaging expanded-beam optics by Moslehi *et al.* (1989). The major consideration is the coupling efficiency, so non-imaging optics are relevant. Several fabrication methods have been tested (see Figure 2.5).

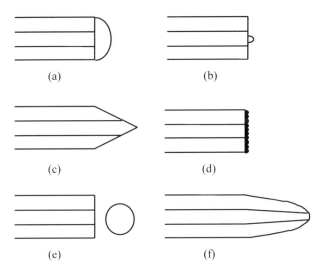

Figure 2.4. Microlenses used in coupling: (a) (b) hemispherical lens, (c) conical lens, (d) Fresnel lens, (e) microsphere, (f) tapered hemispherical fibre end.

In a method for simple and reproducible manufacturing of fibre (or taper) with spherical lenses, with a lens radius of 10 μm, glass is melted in a small container and the plane fibre or taper end is briefly dipped into the melt (see Figure 2.5a). The lens material is dense flint glass with a refractive index $n = 1.8$ at $\lambda = 1.3$ μm.

Fabrication of hemispherical or hemicylindrical lenses on optical fibre surfaces is obtained from light-sensitive polymers (photoresist) crosslinked directly (see Figure 2.5b). A thin film of commercially available photoresist is deposited on the front surface and the resist exposed to UV radiation through the fibre core. The unexposed photoresist is removed; a hemispherical structure has been created. By using a slit to project UV straight on to the front surface of the fibre a hemicylindrical lens can be made. A photolithographic technique can be also used.

Hemispherical microlenses can be manufactured on the ends of single-mode optical fibres using an electric arc discharge. A silica rod with a diameter of about 30–50 μm is placed opposite a single-mode fibre flat end surface (see Figure 2.5c). The electric arc discharge fuses the rod and the fibre end. As the rod is pulled apart, a rounded tip is left on the end of the fibre.

By chemical etching of a single mode fibre in a solution which consists of one part 40% hydrofluoric acid buffered in 10 parts of 40% ammonium fluoride in aqueous solution, truncated conical lenses are formed (see Figure 2.5d). The etching rate is lowest in the central part

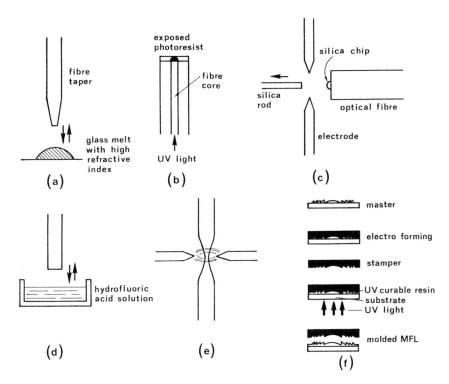

Figure 2.5. Fabrication of integrated microlenses on fibre end faces: (a) dipping the top in molten high index glass [After Khoe et al. 1984, Mathyssek et al. 1985], (b) exposed photoresist [After Cohen and Schneider 1974], (c) fusion of silica rod and fibre end [After Yamada et al. 1980], (d) chemical etching [After Kayoun et al. 1981, Eiseinstein and Vitello 1982], (e) drawing in arc discharge [After Kuwahara et al. 1980], (f) process from a master MFL (moulded fibre lens) to a production MFL [After Tanigami et al. 1989].

of the core where the Ge concentration is highest. The lens has the correct core diameter dimension, is aligned with the core, is easy to manufacture and readily reproducible.

A microlens at a tapered fibre ouput is also obtained by drawing it in an arc discharge (see Figure 2.5e). The hemispherical fibre tip is formed by the surface tension of the molten glass. The radius of curvature is 15–25 mm. In the tapered area the core and cladding thickness decrease by the same ratio. A typical separation between the two electrodes is 2 mm and the discharge time 3 seconds.

In a similar method the single-mode fibre is locally heated in the discharge under tension between a fixed clamp and a moving clamp. The constricted area is asymmetric and the short taper is broken with

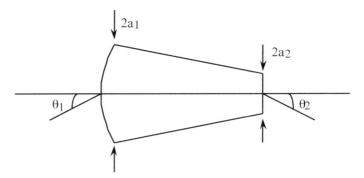

Figure 2.6. Hyperbolic profiled NIO concentrator with end diameters $2a_1$ and $2a_2$. A beam of diameter φ_1 with a divergence angle θ_1 is concentrated into a beam of diameter φ_2 with a divergence angle θ_2 [After Moleshi et al. 1989].

a special tool. The lens is made by dipping the top of the cleaved taper in molten high index glass. The taper is lifted out immediately after contact with the molten glass.

In Figure 2.5f a moulded micro-Fresnel lens (MFL) fabrication process is shown. By electron beam lithography, a master lens is drawn and a stamper is made by a nickel electro-forming method. A UV curable resin is added between the stamper and a glass substrate. Photopolymerisation creates a moulded replica lens. Aberrations are very low.

Hemi-ellipsoidal microlenses have been made by polishing fibre ends into quadrangular pyramids with rectangular cross sections. Hemi-ellipsoids are formed by subjecting these ends to heat fusion from an arc discharge. In coupling experiments between a single mode fibre and a laser diode, improvements of 5.6–9.7 dB have been achieved in comparison with a flat ended single-mode fibre.

Molded aspheric glass lenses have been developed and used in cameras and optical disk drivers. These lenses can be designed by a ray tracing method. They have the advantages of low aberration, low cost and high reliability (Kato and Nishi 1990, Kikuchi et al. 1981). Numerical computation shows that a high degree of aberration compensation has been achieved. To couple sources to multimode fibres, calculations using ray tracing show that maximum coupling efficiency is obtained with one or two spherical lenses of high index material (see Figure 2.4e) (Presby et al. 1988a, Lambert and Goure 1985).

It is possible to fabricate non-imaging optics (NIO) expanded-beam multimode fibre-optics couplers using injection moulding technology. Such plastics couplers and connectors have been packaged in biconic and SMA housings (see Figure 2.6). They could easily be mass produced at low cost.

2.2.3 Tapers

The use of single-mode fibres requires reliable hardware such as field connectors, laser-fibre couplers and directional couplers. The fabrication of these components is difficult because of the small core size of single-mode fibres. The addition of self-aligned beam expansion can eliminate the need for lenses that require critical and stable alignment. The associated single-mode hardware construction is very sensitive to axial and transversal displacements induced mechanically or thermally, as well as to tiny dust particles. The introduction of beam expansion optics alleviates these problems. Conical fibre tapers are fibre sections whose diameter changes continuously along their axis (see Figure 2.7). Fibre tapers can be made very simply by heating a part of fibre in a flame and applying axial tension (see § 2.2.2). By breaking this biconical taper at the waist, one obtains two conical fibre tapers. In a contracting taper, the core radius and the normalized frequency decreases while the spot size increases. The beam cannot widen faster than it does due to diffraction during propagation in a homogeneous medium. Thus, in a strongly contracting taper a great part of guided energy is converted into radiated energy. Practical fibres have a finite cladding surrounded by a protective sheath. During the heating and drawing taper fabrication process, the sheath is removed and the cladding is exposed. Accordingly, tapers are modelled assuming a finite cladding surrounded by air.

Theoretical studies of single mode tapers have been developed: mode conversion using scalar approximation (Marcuse 1987), computer simulation employing a beam propagation method (Amitay and Presby 1989), spot size variation (Love 1987, Henry and Love 1989, Love 1989).

Fabrication, design limitation and characteristics of a single mode tapered beam expander are reported by Keil *et al.* (1984), Jedrzejewski *et al.*(1986), Martinez *et al.* (1988), Kiwahara *et al.* (1996), Martinez *et al.* (1988), Kihara *et al.* (1996). Let us consider a single mode tapered fibre with a step profile (see Figure 2.7e).

The core radius $a_0(z)$ decreases along the z-axis and the ratio of cladding radius $a_1(z)$ to core radius $a_0(z)$ remains constant.

$$a_1(z)/a_0(z) = a_1(0)/a_0(0) \qquad (2.4)$$

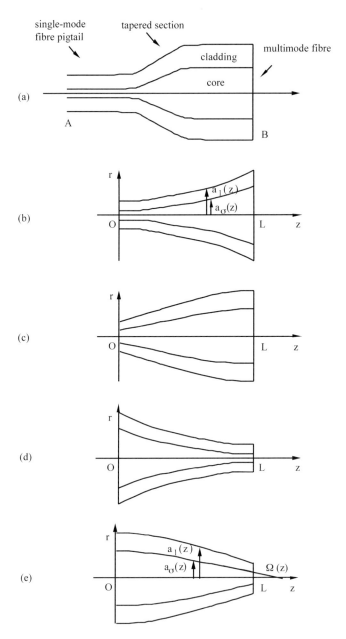

Figure 2.7. Tapered step profile: (a) (b) (c) expansion beam taper, (d) (e) reduction beam taper.

The local V value for a fibre is given by Equation 1.24 and for this kind of taper we have:

$$V(z) = V(0)a_0(z)/a_0(0) \qquad (2.5)$$

If the angle is sufficiently small to ensure negligible coupling between the fundamental and higher order modes, the taper is said to be *adiabatic*. In this case all the energy remains in the fundamental mode. However, because of loss of cylindrical symmetry there is coupling from the fundamental mode into higher-order modes, which are cladding modes. If the taper angle is not small enough, the coupling energy loss is not negligible and the taper is non-adiabatic or lossy.

Consider an infinite adiabatic taper with a parabolic profile defined at a length z from the input by:

$$n^2(R) = n_0^2(1 - 2\Delta R^2) \qquad (2.6)$$

where $R(= r/a_0)$ is the normalized radial distance using the Petermann definition of spot size (Petermann 1977). The local spot size $\omega(z)$ normalized to the initial radius $a_0(0)$ is given by Henry and Love 1989:

$$\omega(z)/a_0(0) = [\omega(0)/a_0(0)][V(z)/V(0)]^{1/2} \qquad (2.7)$$

It increases linearly with increasing $a(z)$ or decreasing taper ratio. This result is different from that of the spot size for a finite clad step-profile taper (see Figure 2.8).

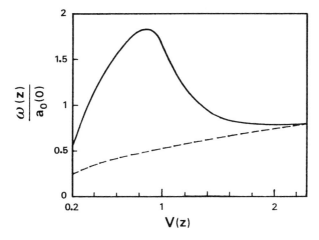

Figure 2.8. Normalized spot size as a function of local V value along a taper for: (a) finite clad, step-profile fibre (solid line), (b) infinite parabolic profile (broken line) [After Henry and Love 1989].

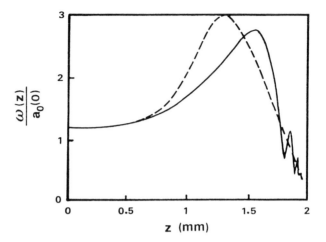

Figure 2.9. Normalized spot size as a function of distance along a linearly decreasing, finite clad, step-profile taper, showing: (a) adiabatic variation (broken line), (b) non adiabatic variation (solid line) [After Henry and Love 1989].

In the case of non adiabatic tapers the spot size is determined by the total field. It results from the superposition of the mode fields over the core cladding cross-section, fundamental mode and higher order cladding modes. Owing to axial symmetry, the HE_{11} local mode is coupled to higher order HE_{1m} modes ($m = 2, 3 \ldots$). The results are more complicated. In the case of finite clad step profile non-adiabatic taper with a linear taper shape we have:

$$a_0(z) = a_0(0)[1 - (z/L)] \qquad (2.8)$$

where L is the length of the taper. The ratio $\omega(z)/a_0(0)$ in the case of adiabatic variation presents a peak at a distance smaller than in the case of non-adiabatic variation (see Figure 2.9).

Calculation of the actual loss is usually involved, and requires sophisticated mathematics (Love (1989), Love et al. 1991, Black et al. 1991). Consider a single-mode fibre whose fundamental mode is excited. Fundamental mode power is coupled to the radiation field by any non-uniformity that alters the cylindrical symmetry. If the single mode is tapered, the cylindrical symmetry is destroyed. In the case of a finite clad fibre the power is lost through coupling between the fundamental mode and cladding modes. The coupling length L_c is defined in terms of the propagation constants β_1 of the fundamental mode HE_{11} and β_2 of the cladding mode:

$$L_c = 2\pi/(\beta_1 - \beta_2) \qquad (2.9)$$

In the case of taper (see Figure 2.7e) a length scale L' is defined in terms of the angle $\Omega(z)$ between the tangent to core cladding interface and the fibre axis. If $a(z)$ is the local radius we have:

$$\tan \Omega(z) = a(z)/L' \qquad (2.10)$$

where L' is the distance along the taper axis from to apex of the cone with half angle $\Omega(Z)$. Scattering theory indicates that if the length scale L' of the non-uniformity is longer than the wavelength of the electromagnetic field, losses or scattering are observed, because the length scale of the field is small enough to detect the non-uniformity. If $n_1 \cong n_2 \cong n$ (weakly guiding approximation), losses are significant when $(\lambda/n < L')$. Significant loss will occur if the length scale of the non-uniformity (taper) is bounded approximately by the scattering and coupling length scale i.e. Love (1989):

$$(\lambda/n) < L' < L_c \qquad (2.11)$$

Thus for an adiabatic taper we have $\tan \Omega(z) \cong \Omega(z)$ and equation (2.9) indicates that if $L' > L_c$ we have low or equivalent loss if:

$$\Omega(z) < a(\beta_1 - \beta_2)/2\pi \qquad (2.12)$$

The taper angle must be sufficiently small and must satisfy the equation (2.12) at each position along the taper.

The basic property of a taper is also to expand the single-mode spot size. These devices have for example a standard single-mode geometry at one end and increase gradually in cross section so that the core size at the other end is comparable to a multimode fibre one. For a sufficiently slow taper the wavefronts remain nearly practically plane. For a gradual taper, which can support multimode propagation at the enlarged end, the conversion of the fundamental mode into higher-order or radiation modes must be negligible if a very low excess coupling loss is maintained. These tapers can be coupled with very low excess-loss (less than 0.1 dB) and are essentially lossless for close coupling.

The field in the fibre taper can be described by coupled mode theory (Snyder and Love 1983, Marcuse 1987, Winn and Harris 1975, or can be determined by solving the scalar wave equation using generalized Laguerre Gaussian polynomials (Bolle and Lundgren 1990), or by employing the beam propagation method (BPM) (Amitay and Presby 1989).

Let us consider coupled mode theory principle as an example. A tapered fibre core radius is increased linearly along z. At each point along z, the cladding to core radius ratio is constant. The refractive index is given by equation (2.1), in which r is a function of z. Using the scalar approximation, the field ψ which can be regarded as the most important transverse component of the electric field, is a solution of the wave equation in cylindrical coordinates:

$$\frac{\partial^2 \psi}{\partial r^2} + \frac{1}{r}\frac{\partial \psi}{\partial r} + \frac{\partial^2 \psi}{\partial z^2} + \frac{1}{r^2}\frac{\partial^2 \psi}{\partial \varphi^2} + n^2(r,z)k_0^2 = 0 \quad (2.13)$$

At any point the field is expressed as a superposition of modes φ_v:

$$\psi = \sum_{v=0}^{\infty} C_v(z)\varphi_v \exp\left[-i\int_0^z \beta_v(z')\,dz'\right] \quad (2.14)$$

The modes φ_v are solutions of the reduced wave equation local normal modes (Marcuse 1975).

$$\frac{\partial^2 \varphi_v}{\partial r^2} + \frac{1}{r}\frac{\partial \varphi_v}{\partial r} + \left[n^2(r,z)k^2 - \beta_v^2\right]\varphi_v = 0 \quad (2.15)$$

The expansion coefficients $C_v(z)$ satisfy the coupled wave equations:

$$\frac{dC_v(z)}{dz} = \sum_\mu R_{v\mu} C_\mu(z) \exp\left[i\int_0^z (\beta_v - \beta_\mu)\,dz'\right] \quad (2.16)$$

$R_{v\mu}$ are the coupling coefficients.

Values of $R_{v\mu}$ and C_μ can be calculated in the case of step index fibre and in the case of a fibre with an infinitely extended radial parabolic index profile. Knowing the coupling coefficients, the coupled wave equation (2.14) can be solved numerically. The results show that the dominant mode adapts itself adiabatically to the changing fibre radius if the change is gradual.

However a difference in the mode power appears between the step-index fibre and the parabolic-index fibre. It is caused by the different dependence of the functions β_0 and β_1 (see Figure 2.10a and b). In the case of a linear taper with a parabolic-index medium the loss curve of mode 0 as a function of length has deep nulls as well as the power of mode 1 (see Figure 2.11c). This work has demonstrated that the fundamental mode in short optical up-tapers ($\sim$1 cm) can be propagated with an insertion loss below 0.1 dB.

The presence of spot size oscillations along a taper is due to a mismatch between the exciting beam at the taper entrance and the fundamental mode. However, slight mismatch should not affect the coupling to a single-mode fibre through such a taper.

For various 1.5 cm long optical up-tapers followed by a straight tip, beam expansion ratios in the range 5–10 are feasible with a corresponding insertion loss per taper of less than 0.01–0.025 dB. In

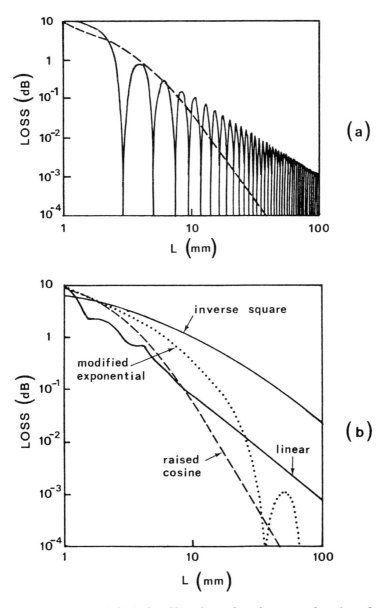

Figure 2.10. (a) Parabolic-index fibre: loss of mode 0 as a function of taper length L of a linear taper (solid line) and for raised cosine tapers (broken line), initial core radius $a_1 = 4$ μm final core radius $a_2 = 361$ μm, (b) Loss of mode 0 as a function of length L for several step index tapers, initial core radius 4 μm, final core radius 48 μm. [After Marcuse 1987].

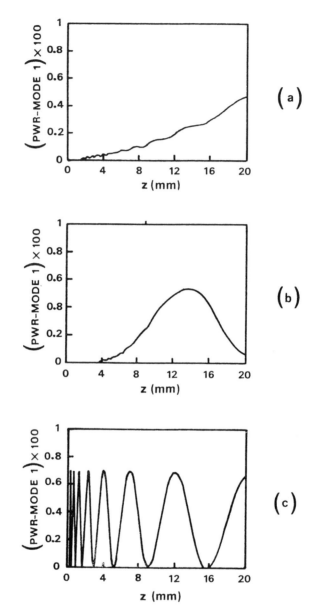

Figure 2.11. Power in mode 1 as a function of z: (a) for step index fibres, linear taper, (b) for step index fibres, raised cosine taper, (c) for a linear taper in a parabolic index fibre [After Marcuse 1987].

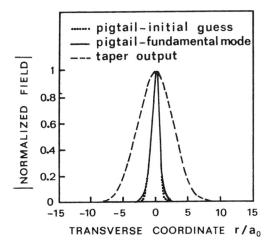

Figure 2.12. Parabolic index taper simulation. Normalized field magnetic versus transverse coordinate r/a_0. $\Delta = 0.3335\%$, $\lambda = 1,3$ μm, $a_0 = 5$ μm, $z = 2$ cm, $\alpha = 20$ [After Amitay and Presby 1989].

the case of parabolic index taper with BPM simulation method, the shape of the fields at different cross sections of the straight tip does not vary, indicating a pure fundamental mode. The beam spot radii at the output are enlarged. However, after a geometrical magnification α of 20 the fundamental mode in the straight tip is confined well within the enlarged core (see Figure 2.12). In the case of step-index tapers the influence of length is important in the mode conversion. The shorter the taper, the bigger is the fundamental mode field deviation, due to higher-order mode conversion (see Figure 2.13). The fundamental mode losses decrease with length (0.36, 0.06 and 0.01 dB for 9, 12 and 24 mm taper lengths).

The insertion loss increases with increasing beam expansion ratio and decreases by increasing the taper length for a constant beam expansion ratio. To couple energy from fibre A to fibre B with very low excess loss (less than 0.1 dB at $\lambda = 0.63$ μm), the use of two similar tapers has the advantage of relaxing the tolerances of the axial and lateral displacements. The excess coupling loss which is defined as the ratio between the output power P_o from the first taper A to the incident power P_e in the second B is expressed by (Amitay et al. 1987):

for axial displacement z

$$T_z = 4/[4 + (z/\lambda)^2/\pi^2(\omega_0/\lambda)^4]$$

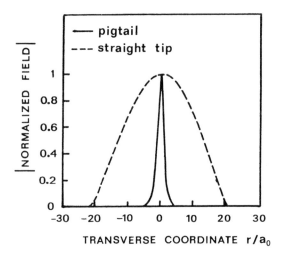

Figure 2.13. Step index taper simulation. Normalized straight tip and fibre pigtail fundamental mode field magnitude versus transverse coordinate r/a_0. $\Delta = 0.5\%$, $n = 1.45$, $\lambda = 1.3$ μm, $\alpha = 20$, $a_0 = 2.41$ μm [After Amitay and Presby 1989].

for lateral displacement d

$$T_d = \exp[-(d/\omega_0)^2] \qquad (2.17)$$

For angular displacement θ

$$T_\theta = \exp[-(\pi \omega_0 \theta/\lambda)^2]$$

with Gaussian beam representation for the dominant modes in the tapers; ω_0 is the equivalent Gaussian beam waist radius for a parabolic profile:

$$\omega_0 = [(a\lambda/n_0\, 2\,\pi)(2/\Delta)^{1/2}]^{1/2} \qquad (2.18)$$

So the sensitivity of the excess loss for lateral and axial displacements of two coupled tapered sections is greatly reduced compared to two coupled single-mode fibres.

For example, for an excess loss of 0.5 dB, the maximum allowed lateral displacement is 3.1 μm for taper coupling (compared to 0.73 μm for fibre coupling). An axial displacement of 291 μm for taper coupling produces 0.5 dB loss (compared with 16.5 μm for fibre coupling). The conversion of the fundamental mode into higher order modes by the taper, which can support multimode propagation at the enlarged end, must be negligible for a gradual taper. These tapers are self-aligning elements.

Tapered single-mode optical fibres can be used in solid state devices using the evanescent field in the tapered area. Coating layers such as sol-gel or polyvinyl acetate solution containing fluorescent materials, permit use of these tapers for sensing applications or amplifiers and laser devices (Henry and Payne 1995).

Recently supercontinuum light with a spectrum 370–1545 nm was generated by femtosecond pulses in a tapered fibre, with potential applications in metrology (Birks *et al.* 2000).

2.3 COUPLING FROM FIBRE TO FIBRE

2.3.1 Alignment losses, butt coupling

2.3.1.1 Splice of circular fields

Common connectors for optical fibres simply bring the fibre ends into direct contact (butt coupling). Low loss splices between optical fibres are indispensable for achieving optical fibre transmittance systems. Splicing is a most important problem for single-mode fibres. Alignment losses are due to lateral, longitudinal and angular offset (see Figure 2.14). Lateral offset is the largest cause of insertion loss

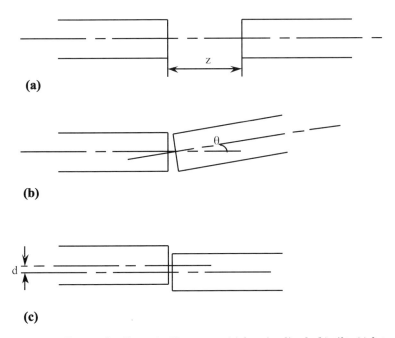

Figure 2.14. Types of splice misalignment: (a) longitudinal, (b) tilt, (c) lateral offset.

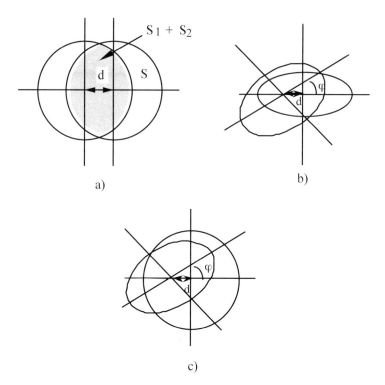

Figure 2.15. Misalignment — transverse offset: (a) of two circular sections, (b) of two elliptic sections, (c) of circular and elliptic section.

(Di Vita and Vannucci 1975, Gordon et al. 1977, Verrier and Goure 1987, Sakai and Kimura 1978).

Let us consider the case of two identical step-index fibres carrying a beam of incoherent light. The relative transmitted power Λ is the ratio of the common area $S_1 + S_2$, divided by the area S of each core (see Figure 2.15a).

$$\Lambda = -10 \, \log_{10}[(S_1 + S_2)/S] \tag{2.19}$$

Multimode fibre

For step-index fibres, if a is the core radius, the coupling efficiency η of a joint between two multimode glass fibres with a lateral offset d is given by Neumann and Weidhaas 1976 as:

$$\eta = (2/\pi) \, \text{arc} \, \cos(\rho/2) - (\rho/\pi)[1 - (\rho/2)^2]^{1/2}$$

where $\rho = d/a$ and for a small offset:
$$\eta \cong 1 - (2\rho/\pi)$$
The resulting coupling loss expressed in decibels is for small offset values:
$$\Lambda(\text{dB}) = -10\ \log_{10}\ [1 - (2\rho/\pi)] \tag{2.20}$$
For a parabolic index fibre
$$\eta = (2/\pi)\ \text{arc}\ \cos(\rho/2) - (\rho/\pi)[4 - \rho^2]^{1/2}(1 - [2 + \rho^2]/12)$$
and for small offsets the coupling loss is
$$\Lambda(\text{dB}) = -10\ \log\ [1 - (4/3)(2\rho/\pi)] \tag{2.21}$$

The insertion loss for longitudinal offset is caused by the divergence of the light beam from the emitting fibre. It is generally calculated by considering the ratio of illuminated area to core area of the second fibre. Under such conditions calculated losses are much larger than those observed. In fact, the irradiance shows a peak on the central axis within a short distance from the emitting area.

Using W_o for power accepted and guided by the second fibre and W_i for the power emitted by the first fibre, the coupling efficiency $\eta = W_o/W_i$ is calculated (Di Vita and Vannucci 1976) utilizing the relation for $W_{o,i}$ as follows:
$$W_{o,i} = 4\pi \int_0^a r\ dr \int_0^{\theta_M} R \sin\theta \cos\theta\ d\theta \tag{2.22}$$
where a is the fibre radius, R the radiance and r represents a point on the fibre end face. θ_M is the maximum value of the angle θ for a guided ray for W_o, $\theta_M = \pi/2$ for W_i and θ is the angle a ray makes with the z-axis. For small separations and for a step index and weakly guiding fibre, the coupling efficiency η and the insertion loss Λ are (Di Vita and Rossi, 1978):
$$\eta = 1 - [(4z\ \text{NA})/(3\pi\ a\ n_0)]$$
$$\Lambda(\text{dB}) = -10\ \log_{10}\ [1 - (4z\ \text{NA})/(3\pi\ a\ n_0)] \tag{2.23}$$
z being the longitudinal offset and n_0 the refractive index of the medium between the fibre ends. If air separates the two fibre ends, a loss of 0.3 dB must be added because of Fresnel reflections at both glass / air interfaces.

The cross-mechanism for connections between graded-index fibres is more complex than for step-index fibres. The modal distribution in the fibre is important. At a short distance from the source, full excitation of all modes is usual; but after a long distance (nearly 1 km) the equilibrium modal distribution (EMD) has taken over, and the modes are more strongly concentrated into the core centre.

For a parabolic index, weakly guiding fibre ($z\,\mathrm{NA}/a \ll 1$), the coupling efficiency η and the insertion loss Λ are given by (Di Vita and Rossi 1978):

$$\eta = 1 - (z\,\mathrm{NA}/2a)$$
$$\Lambda(\mathrm{dB}) = -10\,\log_{10}\,[1 - (z\,\mathrm{NA})/(2a)] \qquad (2.24)$$

Single-mode fibre

Losses caused by the misalignment of two single-mode fibres joined in a splice are easily obtained if the mode is effectively Gaussian in shape. A longitudinal fibre separation Z of two identical fibres gives a power loss (see Figure 2.14a) (Marcuse 1976):

$$\Lambda(\mathrm{dB}) = -10\,\log_{10}\,\{[1 + 4z^2]/[(2z^2 + 1)^2 + z^2]\} \qquad (2.25)$$

where $z = k\,n_2\,\omega_0^2\,Z$ and ω_0 is the width radius of the Gaussian field. n_2 is the refractive index of the cladding.

For a splice with an offset d (see Figure 2.14c):

$$\Lambda(\mathrm{dB}) = -10\,\log_{10}\,[\exp(d^2/\omega_0^2)] = 4.34\,d^2/\omega_0^2 \qquad (2.26)$$

and for a splice with tilt θ (Figure 2.14b):

$$\Lambda(\mathrm{dB}) = -10\,\log_{10}\,\{\exp[-(n_2\,\pi\,\theta\,\omega_0/\lambda)^2]\} = 4.34(n_2\,\pi\,\theta\,\omega_0/\lambda)^2 \qquad (2.27)$$

Theoretically the characteristics of the junction between two optical dielectric waveguides can be calculated by using the Huyghens–Kirchhoff diffraction model (Neumann and Opielka 1977). The splice losses caused by transverse and angular misalignments between two graded-index single-mode fibres can also be evaluated by an efficient theoretical model using the weighted sum of simple Laguerre Gaussian functions (Meunier *et al.* 1991, Meunier *et al.* 1994).

When transverse offset and angular misalignment are present, the loss arising in a joint between single-mode fibres depends on the relative directions of the tilt and the plane of polarisation. The individual losses are additive only when the defects are small (Gambling *et al.* 1978).

Mc Cartney *et al.* (1984) have measured loss histograms for splices between shifted zero dispersion monomode fibres at $\lambda = 1.55\,\mu$m. They have used fusion and gluing techniques. They have found that glue splicing may be required for triangular profile fibre. A low-loss compact optical directional coupler for single mode fibres has been made from two sapphire ball lenses, a beamsplitter and two ferrules (Masuda and

Iwama 1982a, Masuda and Iwama 1982b). The excess loss is 1.5–2.5 dB in the 1.2–1.5 μm wavelength range.

There is often a need for durable fibre splice. Both fusion and mechanical splices are used. The latter require more care. In the fusion-splicing process, the two fibres are carefully aligned and then joined by arc-welding. This process utilizes the surface tension of glass for self-alignment of the two fibres. Many splicers operate on the basis of core alignment, joining the fibres when the lowest insertion loss is observed. This can be detected by observation of the optical power radiated from a sharp bend after the joint. Mechanical splices use V-grooves and index-matched optical cement. This technique is simpler than that of fusion splicing. In an *elastomeric splice*, the V-groove and the opposite lid are made of polyester. The fibre is put in the triangular cavity. A small lateral pressure produces precise alignment. The problems of end separation and losses are reduced by adding index matching fluid. Typical splice loss is 0.1 – 0.2 dB.

2.3.1.2 Splice of elliptic fields

Some work on mode field radius (MFR) definitions of an elliptical field in a single-mode fibre, and the relation with splice losses has been published (Sakar *et al.* 1984, Miller 1986, Marcuse 1976). In the case of non-Gaussian field distribution, a good result is obtained if the width spot size parameters of each direction are defined by the field moments proposed by Ohashi *et al.* 1987.

From the theoretical analysis of a splice between different Gaussian elliptical fields in single-mode fibres, splice losses under various mechanical misalignment conditions have been derived. As an example, let us consider an azimuthal rotation φ (see Figure 2.15b). The transmitting and receiving fields are represented by:

$$E_i(x_i, y_i) = \exp[-(x_i^2/\omega_{xi}^2) - (y_i^2/\omega_{yi}^2)]$$

ω_{xi} and ω_{yi} being the mode field radius along x_i (major) and y_i (minor) axes for each fibre. The splice loss is (Fan and Liang 1990):

$$\Lambda_\varphi = -10 \log_{10}[4/(\eta + \chi \sin^2 \varphi)] \qquad (2.28)$$

where $\chi = [\xi_1 - (1/\xi_1)][\xi_2 - (1/\xi_2)]$ and $\xi_1 = \omega_{x1}/\omega_{y1}$ $\xi_2 = \omega_{x2}/\omega_{y2}$

and where:

$$\eta = [\eta_x + (1/\eta_x)][\eta_y + (1/\eta_y)]$$

and

$$\eta_x = \omega_{x1}/\omega_{x2} \quad \eta_y = \omega_{y1}/\omega_{y2}$$

One result concerns the case of two optical waveguides where one has a circular Gaussian mode field $\xi = 1$ (see Figure 2.15c). The splice loss for azimuthal rotation φ will be reduced to

$$\Lambda_\varphi = -10 \log_{10}(4/\eta) \qquad (2.29)$$

and can be used to calculate the mode field mismatch loss for an LD or $LiNbO_3$ waveguide to SMF coupling (cf. § 2.5.2).

2.3.2 Coupling with microcomponents

The graded-index(GRIN) rod lens (see § 2.2.1) is useful for coupling between multimode optical fibres. One GRIN lens collimates light from an input fibre and a second focuses it on to an output fibre. The systems using lenses are not strictly image-transferring systems. The light radiated by the input fibre needs to be imaged on to the output fibre, but this is not sufficient to achieve lossless coupling because of the limited numerical aperture. Losses due to lens aberrations also have to be added to the losses due to misalignments or spacing, but these aberration losses can be held to be better than 1 dB . Palais (1980) has analysed losses resulting from lateral and angular misalignments. In a well designed system, the transmitting lens will be large enough to contain all the light transmitted by the fibre. Incorrect length of the GRIN lens causes degraded collimation.

Many authors have proposed, the use of beam expansion optics to alleviate alignment problems (Amitay *et al.* 1987, Benner *et al.* 1990) or of lenses (Nicia 1981). Such devices are described in § 2.2. In the first case two tapers are held within a precision ceramic sleeve. The fundamental mode of the single-mode fibre is adiabatically expanded in the first taper. It leaves the taper as a quasi-gaussian propagating beam crossing the space between the tapers. This beam is injected in to the second taper with a coupling efficiency determined by the overlap between the radiation field and the output taper fundamental mode. The device has a low dependence on polarisation and wavelength; attenuation is greater than 75 dB, so the device can be used as a variable optical attenuator.

In the case of single-mode fibre (SMF) components using GRIN lenses the output from the first SMF is assumed to have Gaussian profile. Provided the transmitting SMF and receiving SMF are identical, and provided the Fresnel reflections are suppressed by using index matching, the field inside the GRIN lens can be calculated using the optical transfer function of the lens. The coupling efficiency, computed from the overlap integral of this field distribution and the receiving SMF, shows that a 1% error in length on a half-pitch GRIN

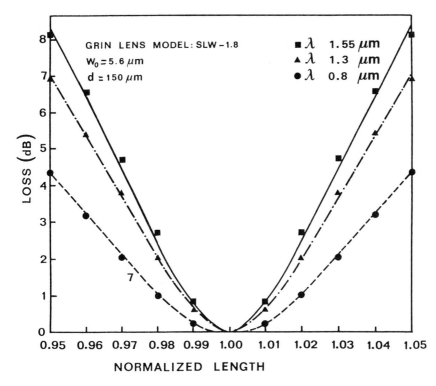

Figure 2.16. GRIN lenses: loss curves for different wavelengths with no misalignment [After Lu et al. 1988].

lens will result in about 0.7 dB loss: the longer the wavelength, the larger is the loss (see Figure 2.16) (Lu et al. 1988).

In the case of angular misalignment of the fibres, there is little loss until the angle is larger than 30° (see Figure 2.17). Two GRIN lenses can be used for coupling bunched optical fibre to single-mode optical fibre with cross-talk attenuation greater than 30 dB (Maekawa and Azuma 1987).

Several other methods of improving the coupling efficiency between a light source and an optical fibre waveguide or between two fibres have been developed using a cylindrical glass fibre as a focusing element (Weidel 1975), a small diameter fibre between the source and the waveguide (Panock et al. 1984), or a thermally expanded core fibre (Kihara et al. 1996).

A process for mass production consists of splicing a micro-optics array on to a ribbon shape at the end face of a single-mode fibre. The micro-optics is composed of two welded sections of a GRIN fibre array and a silica array section (Chanclou et al. 1999).

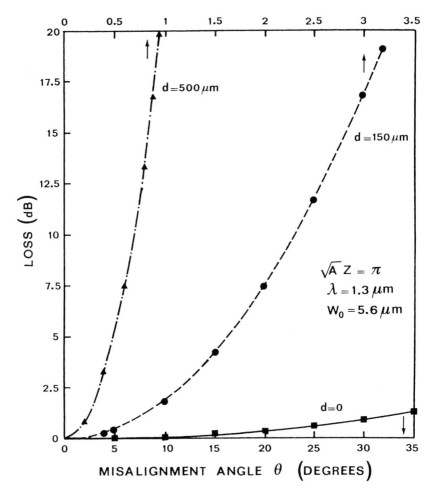

Figure 2.17. GRIN lens: loss curves due to angular misalignment [After Lu et al. 1988].

2.3.3 Non-intrusive tap

The use of devices employing non-invasive or non-intrusive optical fibre taps is of considerable interest, as such devices can be deployed, moved or removed without the need to break the fibre for splicing or connecting. A method of coupling light into and from a fibre is to use a single macrobend of small radius (1 to 5 mm). However, stress in the macrobend makes this method unsuitable because of probable reduction in the life of the device.

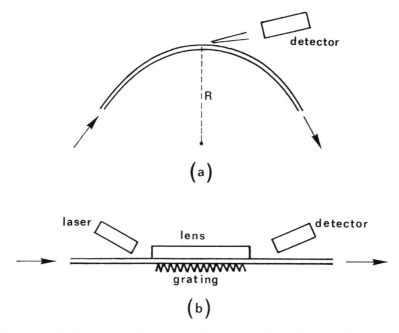

Figure 2.18. Schematic diagram of direct injection/collection fibre tap: (a) single macrobend of radius R, (b) grating and lens [After Cannell *et al.* 1988].

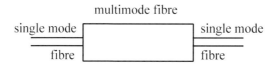

Figure 2.19. Structure of untapered multifibre union [After Horche *et al.* 1989].

The use of a mechanical grating puts much less stress on the fibre. The fibre is sandwiched between the grating and a lens (see Figure 2.18). The grating perturbs the fibre so that it induces coupling between the core and cladding modes. The power is transmitted across the boundary between the cladding and the protective coating. The light is collected and focused by a lens onto a detector. Light injection is obtained similarly. The tapped fraction is defined by the force between the grating and the fibre (Cannel *et al.* 1988). Similar effects can be obtained by Bragg gratings written directly into the fibre (see Chapter 3 § 3.3).

A passive device can be made with a multimode fibre spliced between two single-mode fibres (Horche *et al.* 1989) (see Figure 2.19).

The power transmittance varies with the wavelength. The applications are similar to abruptly tapered single-mode fibres, and the advantage lies in low cost. They are applicable to channel separation in wavelength multiplexed optical fibre systems.

2.3.4 Connectors

The fabrication of graded index fibre connectors has reached the level of routine production. Unlike a laser-to-fibre coupler, a connection is a demountable device that should outlast many disconnections and reconnections (see Figure 2.20). The principles of connectors has been described in numerous papers, e.g., Nawata 1980, Minowa *et al.* 1982. Such a device uses lenses (Nicia 1978, Masuda and Iwana 1982, Kanayama *et al.* 1995) or butt coupling (Khoe *et al.* 1984, Khoe 1986).

A connector in which two lenses produce an image of the emitting fibre end upon the receiving end is not always a good solution as it is difficult to have perfect and reproducible alignment. A single-mode connector based on a butt joint between two ferrule-terminated fibre ends is preferable. The sheathing material is stripped to the exact ferrule length and the bare fibre inserted in the central hole of ferrule. The diameter of the hole exceeds the outside diameter of the fibre by less than 1-2 μm. The fibre is glued in position and the end of the ferrule and fibre are polished. The ferrule is assembled into a plug and two plugs are coupled by an adaptor.

In one connector design, in order to reduce the reflected energy, the two fibre ends are cut at an angle and the two Gaussian modes are tilted (Rao and Cook 1986, Kihara *et al.* 1996).

2.4 COUPLING FROM FIBRE TO WAVEGUIDE

Fibre-to-waveguide couplers are essential for coupling optical signals from fibres to integrated optical signal processing circuits. Coupling efficiency depends on the spatial overlap between the transverse fields of the optical modes in the fibre and in the channel waveguide. The problem has been studied for several kinds of channel waveguide material: silica, silicon, InP and $LiNbO_3$. The geometric mismatch between the waveguide core, which tends to have an elongated cross-section and the fibre core, which is circular, is a complicating factor. As for fibre-to-fibre coupling, it is possible to use lenses or butt coupling but in this case the use of V-grooves to locate the fibres is very important, simplifying alignment and allowing repeatable coupling (Bulmer *et al.* 1980, Sugita *et al.* 1993). A review of fibre attachment to guided wave devices is described by Murphy (1988).

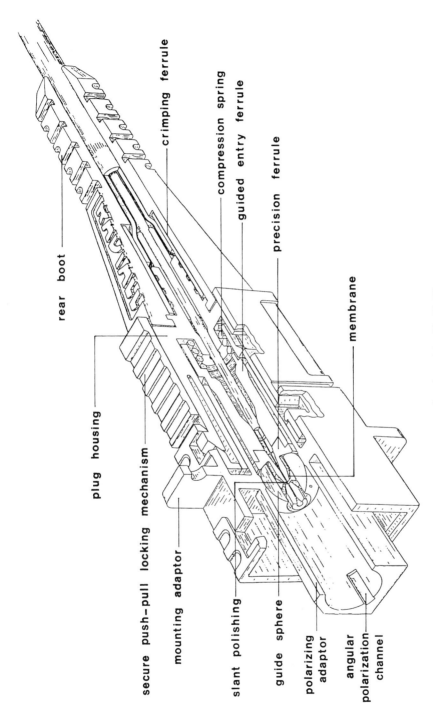

Figure 2.20. Example of connector [Optaball system developed by Radiall].

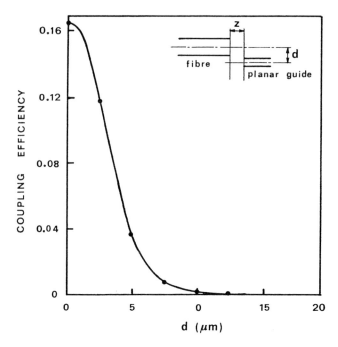

Figure 2.21. Variation of coupling efficiency with lateral displacement d between the two guide axes for a single-mode fibre with $a = 5$ μm, $n_2 = 1.450$, $n_1 = 1.454$ and a dielectric planar guide $2a = 1$ μm, $n_2 = 3.40$ μm and $n_1 = 3.61$ at $\lambda_0 = 1.55$ μm. The interguide distance is $z = 121.8$ μm and $n_0 = 1$ [*After Capsalis and Uzunoglu 1987*].

Calculations of coupling coefficients showing a value of 80–90% order have been made (Burns and Hocker 1977, Ramos et al. 1994, Ramos et al. 1995).

2.4.1 Butt end coupling

In butt end coupling technology both guides are terminated facing one another in contact. A theoretical coupling analysis can be made by a mixed-spectrum eigenwave representations of fields inside the waveguides and Fourier integrals describing the field between the two guides. To satisfy the boundary conditions on the terminal planes of both waveguides, a coupled system of integral equations is derived. The computations show that a coupling efficiency of around 20% could be achieved between the two guides under proper alignment. The most critical tolerance seems to be lateral displacement of the two guides (see Figure 2.21) (Capsalis and Uzunoglu 1987).

Efficient butt-end coupling from fibres to Ti: LiNbO$_3$ channel waveguides has been obtained at 1.32 μm and 1.15 μm (Guttmann et al. 1975, Fukuma and Noda 1980, Ramaswamy et al. 1982, Alferness et al. 1982) as well as for fibres to channel waveguide on glass (Ramos et al. 1994, Rios et al. 1995). Fibre–TiLiNbO$_3$ waveguide coupling loss is minimized by using appropriate diffusion parameters (Mc Caughan and Murphy 1983). The coupling loss is well correlated to the match between the fibre mode diameter and the waveguide geometric mean diameter.

2.4.2 Coupling using grooves

A fibre guiding groove facilitates alignment between fibres and waveguides (LiNbO$_3$ or silica). It has been used for high silica channel waveguides on silicon substrate (Bulmer et al. 1980, Sagita et al. 1993, Nutt et al. 1984). However, alignment of the silicon V-grooves with the waveguide substrate presents a problem. Also the difference in the coefficients of thermal expansion may result in misalignment.

Grooves cut in the same substrate as the waveguide are used for fibre location and alignment (see Figure 2.22.). The grooves may be defined by conventional photolithography and fabricated using ion milling (Murphy and Rice 1986, Nutt et al. 1984, and Bristow et al. 1985) (see Figure 2.23).

A large number of grooves may be cut on one substrate in a single process. Many fibres may be coupled to one waveguide circuit (see Figure 2.24). The advantage of this method is the ease of alignment.

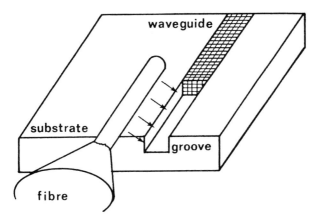

Figure 2.22. Schematic diagram of coupler. Groove is cut in the same substrate as the waveguide [After Nutt et al. 1984].

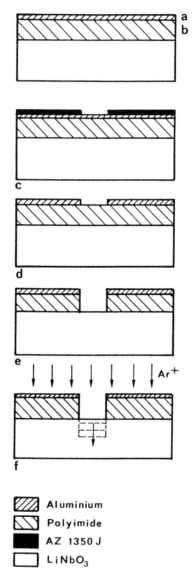

Figure 2.23. Procedure for alignment groove formation in LiNbO$_3$: (a) aluminium layer, (b) polyimide layer, (c) photoresist pattern of groove, (d) resist pattern transferred to Al, (e) reactive ion etched in oxygen, (f) argon ion milling [After Bristow *et al.* 1985].

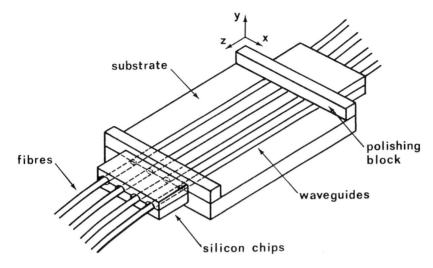

Figure 2.24. Silicon chip coupling [After Murphy *et al.* 1985].

2.4.3 Other methods

An automated method for aligning arrays of fibres to waveguide devices uses the functional dependence of coupling efficiency on fibre waveguide misalignment (Bahadori and Murphy 1989, Murphy *et al.* 1985). If the waveguide substrate ends in an out-of-plane 45° angle, the guided wave will be totally internally reflected out of the waveguide surface through the perpendicular end face. This allows easy connection with a fibre (Kincaid *et al.* 1989).

Low splice loss and zero reflection from the splicing between optical fibre and planar deposited silica waveguide is obtainable by fusion, using a CO_2 laser as a heat source (Shimizu *et al.* 1982).

A ball lens can provide an efficient coupling between a single-mode optical fibre and a silica guide. Theoretical analysis and experiment show that the coupling efficiency is similar to butt-coupling; but tolerances are better with a ball lens (Ramos *et al.* 1995).

Coupling loss theory for a multimode channel waveguide section fabricated by ion exchange in a glass substrate employs a ray-optic approximation (Ctyroky 1984). Coupling losses in the range 2–0.25 dB can be achieved for single channel and sandwiched structures coupled to graded-index fibres.

In an alternative arrangement a single-mode fibre can be efficiently coupled to a single-mode waveguide via a tapered multi-mode area. To couple the optical power efficiently from the fibre without critical matching of the fibre and waveguide modes, a

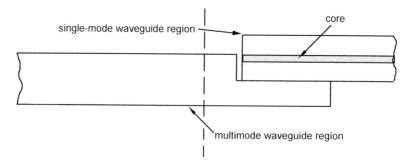

Figure 2.25. Fibre-to-guide coupling via a multimode tapered section [After Chung and Millington 1987].

multimode area that is gradually tapered to a single-mode waveguide is used. With milling, very smooth surfaces can be obtained (see Figure 2.25). The groove milled on the side of a glass substrate is generated by the friction of a micro drill shank in a polishing paste (Chung and Millington 1987).

Another method for coupling signal from single-mode fibres to thin film waveguides (A) employs a coupling waveguide (B) having the same film thickness as the diameter of the fibre. This waveguide B is designed to couple optical power distributively into the second thin film waveguide A with an intermediate layer between them. The two waveguides are parallel. Once incident into waveguide B, optical power will transmit gradually into waveguide A due to the distributive optical coupling between them. The length of waveguide B is approximately $L_c = \pi/\Delta\beta$ where $\Delta\beta = \beta_0 - \beta_1$. β_0 and β_1 are the propagation constants of the 0th and first order mode in the structure (Cai et al. 1991).

To effect a coupling between a thin integrated waveguide (for example InP) and a single-mode fibre, a chip can be constructed by liquid phase epitaxy and wet chemical etching with two masking steps. The chip is then used as a taper device (Pohl et al. 1995).

2.5 COUPLING FROM SEMICONDUCTOR LASER OR LED INTO FIBRES

2.5.1 Multimode fibre

Large diameter optical fibres and fibre bundles are currently used in a growing number of applications. Several papers present an analysis of theoretical limits and practical possibilities of power launching for LEDs (Bodem 1978, Albertin et al. 1974, Lambert and Goure 1985, Di

Coupling from semiconductor laser or LED into fibres 65

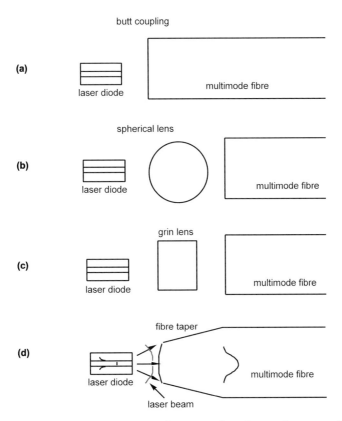

Figure 2.26. Coupling structures from semiconductor laser or LED to multimode fibre.

Vita and Vannucci 1975). Various coupling structures have been presented (see Figure 2.26).

Let us consider the case where multimode fibre and LED are facing each other (see Figure 2.26a). The fibre can support a large number of modes, so the electromagnetic theory of energy launching into the optical fibre becomes very complicated. The energy coupled to an optical multi-mode fibre can be calculated using geometrical analysis. One must consider not only rays that cross the fibre axis (meridional rays), but also those that never meet the axis along their path (skew rays).

Using a cylindrical coordinate system (r, ψ, z) in which the fibre axis represents the z axis (see Figure 2.27), a point (r, ψ), centre of a surface element $dS = r \, dr \, d\psi$ on the source emits rays in a direction defined by the angle θ that they form with the z axis and by the azimuth angle φ with the radius. The total energy W_o emitted by the

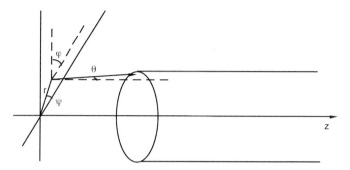

Figure 2.27. Cylindrical coordinates system. Position and angular coordinates of a light ray.

source of area A_s is calculated from its energy $I(r, \psi, \theta, \varphi)$ by integrating the following equation in the emission solid angle:

$$dW_o = I(r, \psi, \theta, \varphi) r \sin\theta \, dr \, d\psi \, d\theta \, d\varphi \qquad (2.30)$$

over θ from 0 to $\pi/2$ (θ being a function of r and φ), over φ from 0 to 2π and over the source area A_s. In the same way the energy W_i collected and guided by the fibre of area A_f is obtained by integrating equation (2.30) over the surface $A_s \cap A_f$, and over θ from 0 to θ_{ca}, where θ_{ca} is the maximum angle for all rays (skew and meridional) corresponding to the fibre numerical aperture (Chapter 1 Equation 1.12). The launching efficiency is:

$$\Lambda = W_i / W_o \qquad (2.31)$$

If the source is an LED emitting uniformly, independently of φ, and having a Lambertian distribution, we have $I(\theta) = I_0 \cos\theta$ and $A_s = \pi b^2$ (b = radius of the LED). So we obtain:

$$W_o = I_0 \pi^2 b^2 \text{ and } W_i = I_0 \pi^2 a^2 \sin^2\theta_{ca}$$

and if

$$A_f < A_s \qquad \Lambda = (a^2 \sin^2\theta_{ca})/b^2 \qquad (2.32)$$

It can be demonstrated that the launching efficiency increases with the fibre acceptance angle.

Coupling efficiency between laser diodes and low loss multimode optical fibres have been calculated using a spherical lens with a high refractive index (> 1.78) and ray tracing (Kawano et al. 1985), and by using a microlens formed at the fibre end (Benson et al. 1975). In the first case the attenuation effects of higher-order modes in the multimode fibre strongly affect the maximum coupling efficiency and misalignment tolerance. In the second case the maximum coupling efficiency achieved by melting the fibre tips to form lens structures varies greatly with fibre taper in the range of 15–63%.

2.5.2 Single-mode fibre

Moreover, in transmitter modules, sensors or instruments, maximum coupling efficiency of the semiconductor laser beam into a single-mode fibre and a low optical feedback into the laser are required. There are several factors that reduce the coupling efficiency of a laser diode to single-mode fibre coupling arrangement. They are ellipticity of the laser diode light, differences in the field shapes and the spot size mismatch of the fields.

The laser diode has an elliptic near field given by the function:

$$\Psi(x,y) = (2/\pi)^{1/2}[1/(\omega_x\omega_y)^{1/2}]\exp\left\{-[(x^2/\omega_x^2) + (y^2/\omega_y^2)]\right\} \quad (2.33)$$

with the assumption of Gaussian field distribution. ω_x is the spot size in the x– direction perpendicular to the junction plane and ω_y is the spot size in the y– direction parallel to the junction plane. For the fibre the near field is given by:

$$\Psi(r) = (2/\pi)^{1/2}[1/\omega]\exp\left\{-(x^2/\omega^2)\right\} \quad (2.34)$$

Various coupling structures have been presented (see Figure 2.28). Butt coupling, lenses, GRIN lenses, spherical microlenses or tapers are used. In the case of coupling devices including lenses, additional causes for the losses are the Fresnel losses, the roughness of the surfaces and the spherical aberration. An increase of coupling efficiency by a factor of 3.5 is obtained in comparison with butt coupling (see Figure 2.28a) when using lenses (Kayoun et al. 1981).

Glass ball lenses and silicon plano-convex lenses are used (see Figure 2.28b,c) (Kartensen and Drogemuller 1990). Lens aberrations, and losses due to misalignment decrease coupling efficiency (Sumida and Takemoto 1984, Kartensen 1988). To reduce the degree of aberration it is necessary to select a lens material of high refractive index and short focal length, in particular for coupling units with ball lenses. Couplers with two lenses have a higher coupling efficiency than couplers with only one lens, and the minimum loss is achieved with lenses of high refractive index and small diameter. Good coupling efficiency has been achieved with a small cylindrical lens (Saruwatari and Nawata 1979).

Coupling can also be made using a moulded aspherical glass lens (see Figure 2.28d) or an aspheric lens with a thermally diffused

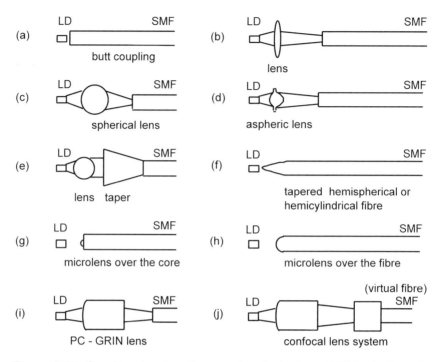

Figure 2.28. Coupling structure from semiconductor laser (LD) to single-mode fibre (SMF).

expanded core single-mode fibre (Kato and Nishi 1990, Kato et al. 1991). Use of taper and ball lens together could be a solution (Figure 2.28e). Several publications report the use of a microlens attached to the single-mode fibre ends (see Figure 2.28g,h) (Gangopadhyay and Sakar 1998). The difficulty is that a rotationally symmetric lens cannot transform an elliptical laser beam into a rotationally symmetric one. The coupling efficiency is improved by using a cylindrical lens. Hemispherical microlenses have been fabricated on the ends of the single-mode optical fibre by using an electric arc discharge (Yamada et al. 1980).

With microscopic lenses, fabricated on optical fibre ends using a photolithographic technique, the coupling efficiency between a laser and single-mode fibre is increased from 8% without a lens to 23% with a spherical lens and 34% with a cylindrical lens (Cohen and Schneider 1974). In another method for increasing the coupling efficiency a high index micro-lens is attached to the cleaved end of a tapered single-mode fibre (Khoe et al. 1983).

GRIN rod lenses (see Figure 2.28i) have been used for collimation and focusing with commercially available optical isolators. The advantage is generous alignment tolerances, typically 40 μm and 250 μm for lateral and axial misalignment (Makita *et al.* 1988).

The drawn fibre taper with a fused spherical microlens on the fibre front end is a coupling arrangement described and tested by Wenke and Zhu 1983. Coupling efficiency of more than 55% is obtained with these coupling structures, with a lens radius of approximately 10 μm.

A tapered waveguide must be considered as a three–layer guiding structure (core–cladding–external medium) and is sensitive to the refractive index of the external medium. If this external medium is a liquid crystal, transmittance should depend on the relative orientation of the liquid crystal director axis with respect to the polarization of the electric field of the fibre. It is possible to use liquid crystals in an electric field, for modulation of side-polished fibre outputs (Veilleux *et al.* 1986). Confocal lens systems can also be used (Figure 2.28j) (Kawahara 1980).

REFERENCES FOR CHAPTER 2

Albertin F, Di vita P and Vannucci R 1974 *Optoelectronics* **6** 369–86
Alferness RC, Ramaswamy VR, Korotky SK, Divino MD and Buhl LL 1982 *IEEE J. Quantum Electron.* **18** 1807–13
Amitay N and Presby HM 1989 *IEEE J. Lightwave Technol.* **7** 131–7
Amitay N, Presby HM, Dimarcello FV and Nelson KT 1987 *IEEE J. Lightwave Technol.* **5** 70–6
Bahadori K and Murphy EJ 1989 *J. Opt. Comm.* **10** 54–5
Benner A, Presby HM and Amitay N 1990 *IEEE J. Lightwave Technol.* **8** 7–10
Benson WW, Pinnow DA and Rich TC, 1975 *Appl. Opt.* **14** 2815–6
Birks TA, Wadsworth W J and Russell P St J 2000 *Opt. Lett.* **25** 1415–17
Black RJ, Lacroix S, Gonthier F, Love JD 1991 *IEE Proc. J.* **138** 355–64
Bodem F 1978 *Optics and Laser Technol* 89–96
Bolle A and Lundgren L 1990 *IEEE Proc. PTJ* **137** 301–4
Bristow JPG, Laybourn PJR, Mc Donach A and Nutt ACG 1985 *IEEE Proc. PTJ* **132** 291–6
Bulmer CH, Sheem SH, Moeller RP and Burns WK 1980 *Appl. Phys. Lett.* **37** 351–55
Burns WK and Hocker GB 1977 *Appl. Opt.* **16** 2048–50
Cai Y, Mizumoto T, Ikegami E and Naito Y 1991 *J. of Lightwave Technol.* **9** 577–83
Cameron KH 1984 *Electron. Lett.* **20** 974–6
Cannell GJ, Epworth RE, Hale PG, King JP, Large T, Leggett CM, Robinson A, Williams RL and Worthington R 1988 *Electron. Lett.* **24** 1534–6

Capsalis CN and Uzunoglu N K 1987 *IEEE Trans Microwave Theory and Techniques* **35** 1043–51
Chanclou P, Thual M, Lostec J, Pavy D, Gadonna M and Poudoulec A 1999 *J. Lightwave Technol.* **17** 924–8
Chen Z, Chen X and Lai H 1992 *IEE Proc. J* **139** 309–12
Chung PS and Millington MJ 1987 *IEEE J. Lightwave Technol.* LT1721–6
Clement DP, Osterberg U and Lasky RC 1993 *IEEE Photonics Technol. Lett.* **5** 1442–4
Cohen LG and Schneider MV 1974 *Appl. Opt.* **13** 89–94
Ctyroky I 1984 *J. Opt. Comm.* **5** 93–9
Di Vita P and Rossi U 1978 *Alta Frequenza* **5** 414–23
Di Vita P and Vannucci R 1975 *Opt. Comm.* **14** 139–44
Di Vita P and Vannucci R 1976 *Appl. optics* **15** 2765–74
Eftimov T and Hitchen P 1993 *Intern Journal of Optoelectronics* **8** 123–32
Eisenstein G and Vitello D 1982 *Appl. Opt.* **21** 3470–4
Fan C and Liang A 1990 *IEEE J. Lightwave Technol.* LT8 173–6
Fukuma M and Noda J 1980 *Appl. Opt.* **19** 591–7
Gambling WA, Matsumura H, Ragdale CM 1978 *Progress in optical communication* (Peter Peregrinus) 162–3
Gangopadhyay S and Sarkar S N 1998 *J Opt Commun* **2** 42–4
Gomez Reino C 1992 *Intern. J. of Optoelectronics* **7** 607–80
Gonthier F, Lapierre J, Veilleux C, Lacroix S and Bures S, 1987 *Appl. Opt.* **26** 444–9
Gordon KS, Rawson EG and Norton RE 1977 *Appl. Opt.* **16** 2372–4
Goure JP, Lambert AM and Massot JN 1984 *Opt. Quant. Elec.* **16** 49–56
Guttmann J, Krumpholz O and Pfeiffer E 1975 *Appl. Opt.* **14** 1225–7
Hasegawa O and Namazu R 1980 *J. Appl. Phys.* **51** 30–6
Henry WM and Love JD 1989 *IEEE Proc. PtJ* **136** 219–24
Henry WM and Payne F P 1995 *Opt. Quant. Elec.* **27** 185–91
Horche PR, Lopez-Amo M, Muriel MA and Martin Pereda JA 1989 *IEEE Photonics Techn. Lett.* **1** 184–7
Ishikawa R *et al.* 1985 *IOOC-ECOC Conference* Venice (Italy)
Jedrzejewski KP, Martinez F, Minelly JD, Hussey CD and Payne FP 1986 *Electron. Lett.* **22** 105–6
Kanayama K, Nagase R, Kato K, Oguchi S, Yoshizawa T and Nagayama A 1995 *IEEE Photonics Technol. Lett.* **7** 520–2
Karstensen H 1988 *J. Opt. Commun.* **2** 42–9
Karstensen H and Drögemüller K 1990 *J. Lightwave Technol* **8** 739–47
Kato K and Nishi I 1990 *IEEE Photonetics Technol. Lett.* **2** 473–4
Kato K, Nishi I, Yoshino K and Hanafusa H 1991 *IEEE Photonics Technology Letters* **3** 469–70

Kawachi M, Yamada Y, Yasu M and Kobagashi M 1985 *Electron. Lett.* **21** 314–15
Kawano K, Miyazawa H and Mitomi O 1985 *Electron. Lett.* **21** 609–11
Kayoun P, Puech C, Papuchon M and Arditty HJ 1981 *Electron. Lett.* **17** 400–2
Keil R, Klement E, Mathyssek K and Wittmann J 1984 *Electron. Lett.* **20** 621–2
Kihara M, Nagasawa S and Tanifuji T 1996 *J. Lightwave Technol.* **14** 542–8
Kihara M, Matsumoto M, Haibara T and Tomita S 1996 *J. Lightwave Technol.* **14** 2209–14
Khoe GD, Van Leest JHFM, Luijendijk JA 1982 *IEEE J. of Quantum Electronic* **18** 1573–80
Khoe GD, Poulissen J and De Vrieze HM 1983 *Electron. Lett.* **19** 205–7
Khoe GD, Kock HG, Küppers D, Poulissen JHFM and DeVrieze HM 1984 *J. Lightwave Technol.* **LT-2** 217–27
Kikuchi K, Morikawa T, Shimada J and Sakurai K 1981 *Appl. Opt.* **20** 388–94
Kincaid BE, Blachman R, Nightingale JL and Becker RA 1989 *Opt. Lett.* **14** 335–7
Kotsas A, Ghafouri-Shiraz H and Mac Lean TSM 1991 *Opt. and Quantum Electron.* **23** 367–78
Kuwahara H, Sasaki H and Tokoyo N 1980 *Appl. Opt.* **19** 2578–83
Lambert AM and Goure JP 1985 *Opt. and Quantum Electron.* **17** 87–90
Love JD 1987 *Electron. Lett.* **23** 993–4
Love JD 1989 *IEEE Proc. PtJ* **136** 225–8
Love JD and Henry WM 1986 *Electron. Lett.* **22** 912–14
Love JD, Henry WM, Stewart WJ, Black RJ, Lacroix S and Gonthier F 1991 *IEE Proc. J* **138** 343–54
Lu Y, Palais JC and Chen Y 1988 *Fiber and Integrated Optics* **7** 85–107
Maekawa E and Azuma Y 1987 *IEEE J. Lightwave Technol.* **LT5** 206–10
Makita Y, Yamauchi I and Sono K 1988 *Fiber and Integrated Optics* **7** 27–33
Marcuse D 1975 *Bell System Techn. J* **54** 1507–29
Marcuse D 1976 *Bell System Techn. J* **56** 703–18
Marcuse D 1987 *J. Lightwave Technol.* **LT-5** 125–33
Martinez F, Wylangowski G, Hussey CD and Payne FP 1988 *Electron. Lett.* **24** 14–16
Masuda S and Iwama T 1982a *Appl. Opt.* **21** 3484–8
Masuda S and Iwama T 1982b *Appl. Opt.* **21** 3475–83
Mathyssek K, Wittmann J and Keil R 1985 *J. Opt. Commun.* **6** 142–6
McCartney DJ, Payne DB and Wright JV 1984 *Electron. Lett.* **20** 78–80

McCaughan L and Murphy EJ 1983 *IEEE J. Quantum Electron.* **19** 131–6
Meunier JP and Hosain SI 1991 *J. of Lightwave Technology* **9** 1457–63
Meunier JP, Wang ZH and Hosain SI 1994 *IEEE Photonics Technology Letters* **6** 998–1000
Miller CM 1986 *Optical Fiber Splices and Connectors : Theory and Methods*(Marcel Dekker: New York) 155
Minowa JI, Saruwatari M and Suzuki N 1982 *IEEE J. of Quantum Electron.* **18** 705–17
Moleshi B, Ng J, Kasimoff I and Jannson T 1989 *Opt. Lett.* **14** 1327–9
Murphy EJ 1988 *J. Lightwave Technol.* **6** 862–71
Murphy EJ and Rice TC 1986 *IEEE J. Quantum Electron.* **22** 928–32
Murphy EJ, Rice T C, McCaughan L, Harvey G T and Read P H 1985 *J. Lightwave Technol.* **3** 795–8
Nawata K 1980 IEEE *J. Quant. Electron.* **16** 618–27
Neumann E G and Opielka D 1977 *Opt. Quant. Electr.* **9** 209–22
Neumann EG and Weidhaas W 1976 *Archiv für Electronik und Ubertragungstechnik* **30** 448–50
Nicia A 1978 *Progress in Optical Commun.* (Peter Peregrinus) 163–4
Nicia A 1981 *Appl. Opt.* **20** 3136–45
Nutt ACG, Bristow JPG, McDonach A and Laybourn PJR 1984 *Opt. Lett.* **9** 463–5
Ohashi M, Kuwaki N and Vesugi 1987, *J. Lightwave Technol.* **5** 1676–79
Palais JC 1980 *Appl. Opt.* **19** 2011–18
Panock R, Forrest SR, Kohl PA, De Winter JC, Nahory RE and Yanowski ED 1984 *J. Lightwave Technol.* **LT2** 300–5
Petermann F 1977 *Opt. and Quantum Electron.* **9** 167–75
Pohl A, Fouckhardt H and Unger HG 1995 *J. Opt. Commun.* **16** 138–42
Presby H, Amitay N, Dimarcello F V and Nelson K T 1987 *J. Lightwave Technol.* **5** 1123–8
Presby HM, Amitay N and Benner A 1988b *Electron. Lett.* **24** 34–5
Presby HM, Amitay N, Scotti R and Benner A 1988a *Electron. Lett.* **24** 323–4
Ramaswamy V, Alferness RL and Divito M 1982 *Electron. Lett.* **18** 30–1
Ramos M, Verrier I, Réglat M, Sass P and Goure JP 1994 *J. Opt. Commun.* **15** 190–6
Ramos M, Verrier I, Goure JP and Mottier P 1995 *J. Opt. Commun.* **16** 179–85
Rao R and Cook JS 1986 *Electron. Lett.* **22** 731–2
Rios S, Srivastava R and Gomez-Reino C 1995 *Optics communications* **119** 517–22

Saitoh A, Gotoh T and Tanaka K 2000 *Opt. Lett.* **25** 1759–61
Sakaguchi H, Seki N and Yamamoto S 1981 *Electron. Lett.* **17** 425–6
Sakai J I and Kimura T 1978 *Appl. Opt.* **17** 2848–53
Sakar S, Thyayrajan K and Kumar A 1984 *Opt. Commun.* **49** 178–83
Saruwatari M and Nawata K 1979 *Appl. Opt.* **18** 1847–56
Shimizu N, Imoto N and Ikeda M 1983 *Electron. Lett.* **19** 96–7
Snyder and Love 1983 Optical Waveguide Theory Chapman and Hall (London: New York)
Sugie T and Saruwatari M 1983 *J. Lightwave Technol.* **LT-1** 121–30
Sugita A, Onose K, Ohmori Y and Yasu M 1993 *Fiber and Integrated Optics* **12** 347–54
Sumida M and Takemoto K 1984 *J. Lightwave Technol.* **2** 305–11
Tanigami M, Ogata S, Aoyama S, Yamashita T and Imanaka K 1989 *IEEE Photonics Technol. Lett.* **1** 384–5
Tomlinson W J 1980 *Appl. Opt.* **19** 1117–26
Veilleux C, Lapierre J and Bures J 1986 *Opt. lett.* **11** 733–5
Verrier I and Goure JP 1987 *J. Opt. Commun.* **8** 151–4
Wang Z, Mikkelsen B, Pedersen B, Stubkjaer KE and Olesen DS 1991 *J. of Lightwave Technol.* **9** 49–55
Weidel E 1975 *Electron. Lett.* **11** 436–7
Wenke G and Zhu Y 1983 *Appl. Opt.* **22** 3837–44
Winn RK and Harris JH 1975 *IEEE Trans Microwave Theory Technol.* **MIT 23** 92–7
Yamada JI, Murakami Y, Sakai JI and Kimura T 1980 *IEEE J. Quantum Electron.* **QE 16** 1067–72

EXERCISES

2.1 – What is the length value of a full pitch SELFOC which has a relative refractive index difference $\Delta n/n = 10^{-3}$ and a core diameter $a = 50$ μm?
What is the length value for a quarter pitch SELFOC lens?

2.2 – Calculate the radiance I_0 at the output of a multimode step index optical fibre of radius $a = 50$ μm, $n_1 = 1.46$ and $\theta_{ca} = 17°$ emitting a power of 50 mW if I_0 is independent of θ.

2.3 – A similar fibre is located at 10 μm from the fibre of exercise 2.1 and perfectly aligned on the same axis. What is the power injected into the second fibre?

2.4 – In a connector the fibre ends of two identical step index fibres (core radius = 25 μm) have a lateral offset of 3 μm. What is the resulting coupling loss?

2.5 – A LED (Lambertian source) with a radius $b = 40$ μm and a power 5 mW is used in front of a multimode fibre

    a – calculate the radiance I_0
    b – calculate the power injected and transmitted by the fibre (core radius = 25 µm, NA = 0.21) in the case of butt-coupling
    c – what is the coupling efficiency Λ

2.6 – The light emitted by a point source is injected in an optical fibre of core diameter $d = 100$ µm and of numerical aperture NA = 0.3. What is the distance for a good coupling?

Chapter 3

DEVICES BASED ON COUPLING EFFECT WITH NON-POLARIZED LIGHT

Over the past few years there has been a growing interest in the use of optical fibres for information distribution systems and sensors. Consequently the fabrication of multiport fibre devices to divide or combine transmission signals, called power couplers, has been developed. Fibre couplers are the most important devices in fibre applications such as high-speed data link systems, wavelength demultiplexer–multiplexer systems, coherent transmission systems, fibre sensors and fibre optics measurement systems.

An optical coupler is usually a passive device that distributes power from the main fibre to one or more branch fibres. Figure 3.1 shows a 2×2 coupler. Several techniques have been developed to construct couplers. Each has some advantages and drawbacks, and several types have been proposed using optical fibres and micro-optics components, e.g. polished cladding couplers, fused biconical taper couplers, beamsplitters, micro-bend types, and lateral offset types. A classification of fibre optic power splitters is given by Agarwal 1985.

Several theoretical techniques have been applied to investigate the behaviour of the optical power distribution in the two fibres and the interlay waveguide as one function of the fibre and waveguide propagation parameters. One utilises the beam propagation method (Van Roey *et al.* 1981, Lamouroux and Prades 1987); an alternative approach is to use the coupled mode theory or the effective index method (Qian 1986, Bures *et al.* 1983, Huang and Chang 1990 a, b, Okamoto 1990, Zheng and Snyder 1987, Wright 1986).

Fibre couplers may be categorized into polarization-independent and -dependent couplers. The latter are classified into polarization-maintaining and -splitting couplers, and these will be examined in Chapter 4.

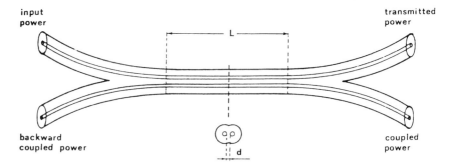

Figure 3.1. X coupler realized from two fibres with a core spacing d. The energy is transferred by an evanescent wave along the interaction length L.

3.1 COUPLING THEORY FOR CIRCULAR FIBRES

Coupling between two single-mode fibres is achieved when the fields associated with their LP_{01} modes interact. In order to access the field of a single-mode fibre we must either taper the fibre down or locally remove a part of its cladding. In a tapered fibre the field width (spot size) increases as the radius of the core is reduced. When two fibres lie side by side in close proximity their field distributions overlap. Access to the field is achieved by grinding and polishing some of the cladding away leaving a thin layer above the core cladding interface, then bringing the polished regions together. The coupling between the fibres takes place through their evanescent fields.

We examine propagation along, and power transfer between, two parallel fibres (see Figure 3.1). The field of a fibre extends indefinitely into the cladding, interacts with the second fibre, and thus excites its field. In turn, the field of the second fibre interacts with the field of the first. As the field propagates there is an exchange of power between the two fibres. This phenomenon is the *optical cross-talk*. The amount of power exchange depends on the amount of overlap of the fields of the two fibres.

Cross-talk can be described in terms of the modes of the composite waveguide, and is manifested by the beating (or interference) of the composite mode; or in terms of the modes that propagate along each fibre independently, when we have sufficiently separated weakly guiding fibres. If there are more than two propagated modes, the first method is very difficult, because it requires a large number of coupled equations. Theoretical explanations can be found in Snyder and Love (1983), Ankiewiz *et al.* (1986), Peng and Ankiewicz (1991) and Huang and Chang (1990 a, b).

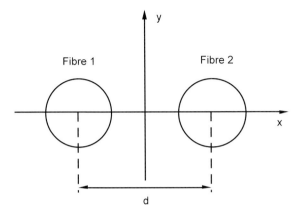

Figure 3.2. Cross section of a fibre optical coupler.

In the first method, let us consider a composite two-fibre waveguide with a cross section as in Figure 3.2. Two identical (or nearly identical) parallel fibres are optically separated. Using these two single-mode fibres in isolation and within the weak guidance approximation, the field of the composite waveguide is approximated by a sum of the fields of the two fibres in isolation.

Each single-mode fibre has two orthogonally polarized fundamental modes. These fields are obtained from the fundamental solution of the scalar wave equation. In the case of two fibres in the composite waveguide, the number of propagated modes is multiplied by two, owing to symmetry. We have two pairs of orthogonally polarized fundamental modes (see Figure 3.3).

The scalar wave equation solutions of the fibre isolation are respectively ψ_1, ψ_2 and the composite waveguide solution is $\psi(x,y)$. The symmetry of the composite waveguide leads to two fundamental solutions for $\psi(x,y)$:

$$\psi_+ = \psi_1 + \psi_2 \qquad \psi_- = \psi_1 - \psi_2 \qquad (3.1)$$

The propagation constants associated with ψ_+ and ψ_- are β_+ and β_- respectively and β is associated with ψ_1 or ψ_2. So the four fundamental modes of the two identical fibres are formed by pairs of antisymmetric and symmetric modes ψ_+ and ψ_-. The transverse electric fields are polarized parallell to the x or y axes. Figure 3.3 shows the orientation of the transverse electric field for the four fundamental modes. The power at the end of the fibres is:

$$P_1(z) = \cos^2(Cz)$$
$$P_2(z) = \sin^2(Cz) \qquad (3.2)$$

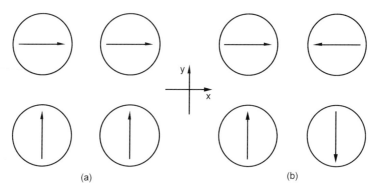

Figure 3.3. (a) Symmetric, (b) antisymmetric modes.

C is the propagation constant due to the coupling modes and is related to the composite waveguide geometry (Snyder and Love 1983). A proportion of the power is transferred from one fibre to the other fibre and back in a beat length:

$$L_b = \pi/C \qquad (3.3)$$

In the second method, using two single-mode fibres in isolation from each other and within the weak guidance approximation, the j-polarized fundamental modes ($j = x$ or y) are

$$E_j(x,y,z) = a_j \Psi_j(x,y) e^{i\beta_j z} = b_j \Psi_j(x,y) \qquad (3.4)$$

with

$$a_1 = 2 \cos Cz$$

$$a_2 = 2i \sin Cz$$

$$b_j = a_j \, e^{i\beta_j z}$$

where β_j is the propagation constant of the mode. The fields of the composite wave guide are approximated by a linear combination of the modes of the fibres in isolation, when the two fibres are well separated and not too dissimilar:

$$\Psi(x,y) = b_1 \Psi_1(x,y) + b_2 \Psi_2(x,y) \qquad (3.5)$$

Two coupled equations are obtained by treating one fibre as one perturbation of the second:

$$db_1/dz - i(\beta_1 + C_{11})b_1 = i\, C_{12}\, b_2 \qquad (3.6)$$

$$db_2/dz - i(\beta_2 + C_{22})b_2 = i\, C_{21}\, b_1$$

where C_{ij} are coupling coefficients, independent of z.

In the case of identical and nearly identical fibres, the power transfer between two modes can be written in terms of the initial power. We can set $C_{12} = C_{21} = C$ and $C_{11}(C_{22}$ being negligible), and eliminate b_1 or b_2 from the equation (3.6) to obtain a second-order differential equation. With unit power launched in the first fibre and zero power in the second, the power transfer between two lossless waveguides in close proximity, having a constant separation d between them along the coupling region (parallel-fibre coupler), is given by:

$$P_1(z) = 1 - F^2 \sin^2[(Cz/F)]$$

$$P_2(z) = F^2 \sin^2[(Cz/F)] \quad (3.7)$$

With

$$F = [1 + (\beta_1 - \beta_2)^2/4\, C^2]^{-1/2}$$

A fraction F^2 of the power is transferred from one fibre to the other fibre and back in the beat length $L_b = \pi F/C$. For two identical fibres $F = 1$.

In these conditions of weak interaction through a tunnel barrier the power will be totally transferred from one guide to the other with a spatial periodicity L given by $L = \pi/2C$.

For two identical step profile fibres, the coupling coefficient (Snyder and Love 1983) becomes:

$$C = (\pi\, \Delta/w\, d\, a)^{1/2}\, (u^2/V^3)[(\exp(-w\, d\, a))/K_1^2 w] \quad (3.8)$$

For each fibre, V, u and w have been defined in equations (1.23) and (1.24). In the case of two identical Gaussian profile fibres, the profile for the composite waveguide is given by:

$$n^2 = n_1^2 \{1 - 2\Delta[1 - \exp(-r_1^2/a^2) - \exp(-r_2^2/a^2)]\} \quad (3.9)$$

r_1 and r_2 are the radial coordinates taken from each fibre centre, n is the fibre core refractive index and the coupling coefficient is:

$$C = (\pi \Delta/d\, a)^{1/2}\, V^3 \left[(V-1)^{1/2}/(V+1)^{-1/2}\right]$$
$$\times \exp\left\{(V-1)[(V-1)/(V+1) - (d/a)]\right\} \quad (3.10)$$

Coupled mode equations can be used for two or more coupled weakly guiding fibres and for double core fibres (Qian 1986).

Using the coupled mode theory in the vectorial form based on the exact HE_{11} and HE_{21} modes, the effect of the polarizations has been considered. Both fundamental modes of single-mode fibres become coupled through their evanescent fields. As explained at the beginning of this chapter, the action of a single mode-fibre coupler relies on the beating of the two lowest order modes. If β_{11}^x (or β_{11}^y) and β_{21}^x (or β_{21}^y) are the propagation constants for the HE_{11}^x (or HE_{11}^y) and HE_{21}^x (HE_{21}^y) modes, the coupler is defined by two polarization-dependent coupling coefficients $Cx = (\beta_{11}^x - \beta_{21}^x)/2$ and $Cy = (\beta_{11}^y - \beta_{21}^y)/2$. HE_{ij}^p is the polarized mode ($p = x$ or y) with i = 1 and j = 1, field zero in the x and y directions (Huang and Chang 1990a).

As an example, in the case of fused tapered single-mode fibre couplers, the fields of the HE_{11}^{xy} and HE_{21}^{xy} modes along the centre line of the coupler ($y = 0$) at various values of $V = b\,k\,(n_2^2 - n_3^2)^{1/2}$ are given in Figure 3.4; $n_3 = 1$ for air and b is the cladding radius. The cladding–core radius ratio $\frac{b}{a}$ and the core separation d have significant effects on the coupling characteristics.

3.2 DIRECTIONAL COUPLERS

3.2.1 X-coupler, 2 × 2 coupler

The X-coupler (also called a 2 × 2 couple) is the guided equivalent of the conventional beam splitter (see Figure 3.1). The purpose is to bring the fibre cores close to each other so that efficient power coupling can take place. The energy is transferred from the excited optical fibre to the adjacent parallel fibre because an appreciable part of the energy is propagated into the cladding by the evanescent wave, as seen in § 3.1. Most coupling schemes therefore try to position the cores as close to each other as possible.

One method of fabricating these couplers consists of polishing one side of the core for both fibres fixed in holders, and then bringing the cores into contact with an index-matching liquid interface (see Figure 3.5).

Two quartz blocks are prepared by first grinding their respective top and bottom faces parallel. A slot is cut into the top face with the bottom of the slot giving a downward curvature. This gives the path of the fibre a curvature which controls the length of the interaction region. The fibre is prepared by stripping the sheath with the help of a solvent for a length equal to that of the holder. The fibres are fixed in the slots. The blocks are polished on one side of the cores in order to partially remove the cladding. The two blocks are then brought into contact. A refractive index matching fluid is inserted by capillary action. The result is a reciprocal power transfer from one fibre to the other. The power transfer ratio depends on the core spacing and on the interaction length. Detailed theoretical and experimental analysis can be found in Bergh et al. (1980), Annovazzi-Lodi and Donati (1990), Leminger and Zengerle (1990).

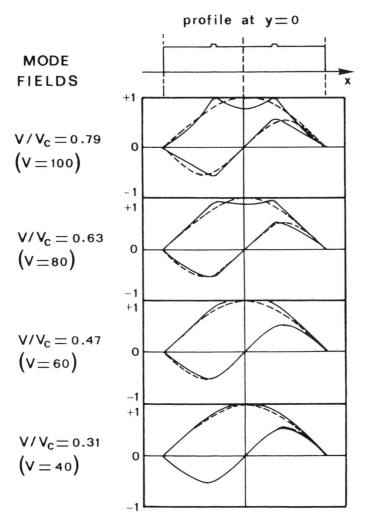

Figure 3.4. Fields of the symmetric HE_{11}^{xy} and antisymmetric HE_{21}^{xy} modes in a coupler with core (solid lines) or without core (broken lines) at various values of V. $d = 1.2\ b$, $\Delta = 0.003$, $b/a = 20$, a and b core and cladding thicknesses [After Chiang 1987].

A similar device is obtained when an optical fibre is put on a low curvature plano convex lens and fixed to it with a thin film of epoxy resin (Parriaux *et al.* 1981a,b).

During the fabrication of the coupler, one of the main factors is the amount of cladding remaining on each substrate. The correct spacing between the surface and the fibre axis can be determined non-

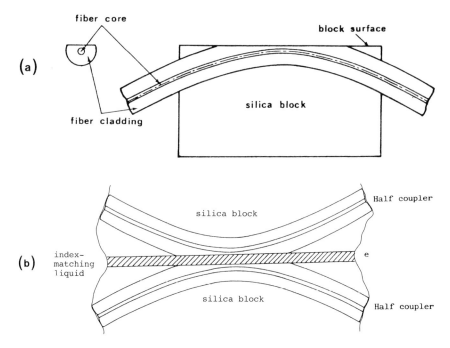

Figure 3.5. Polished single mode optical fibre: (a) half coupler block, (b) scheme of the coupler.

destructively by measurement of the attenuation obtained when a drop of liquid with a known refractive index is placed to cover the whole polished surface (Digonnet et al. 1985, Lamouroux et al. 1985). In a similar method for exact determination of the variable core-to-surface spacing along the polished region of the fibre coupler blocks, a drop of liquid with a known refractive index n_3 slightly higher than the effective index n_e of the fundamental mode is placed on part of the surface. The attenuation is measured while the liquid gradually covers the block surface (Leminger and Zengerle 1987).

An all-fibre device such as the polished fibre coupler could lead to a power transfer P_i, $i = 1, 2$ along the interaction region toward positive z given in Figure 3.6a, relative power division versus wavelength λ given in Figure 3.6b and coupler wavelength selectivity versus λ shown in Figure 3.7. It is possible to see experimentally how the coupling efficiency varies in such a coupler, using a tunable single-mode optical fibre coupler. In a holder provided with micrometer screw the position of the top substrate is adjusted with respect to the bottom substrate, offseting the top fibre by any desired amount. When the spacing of the fibres is increased, the coupling between the fibres decreases and allows fine tunning of the transfer ratio. Figure 3.8 shows experi-

Directional couplers 83

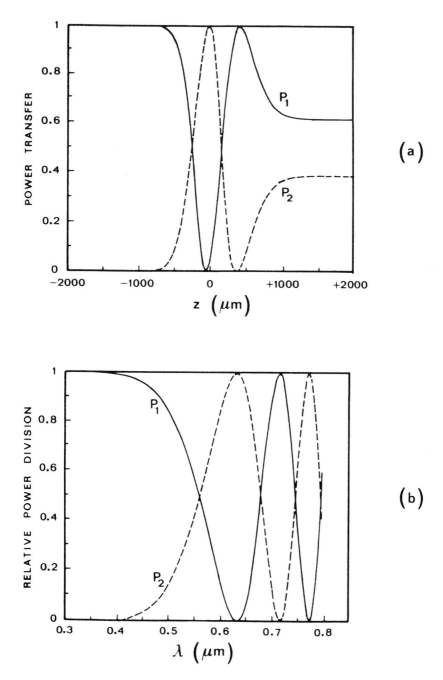

Figure 3.6. (a) Power transfer P_i, i = 1,2, $n_1 = 1.475$, $n_2 = 1.46$, $n_e = 1.47$, $2d = 0.6$ μm, $a = 1.5$ μm, curvature radius $R = 0.5$ m, $\lambda = 0.6328$ μm, (b) Relative power division versus wavelength λ, $d = 0.54$ μm [After Parriaux *et al.* 1981a].

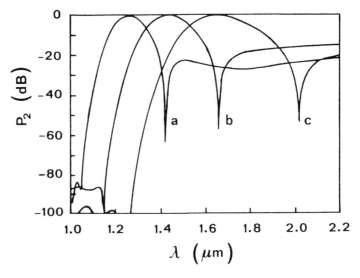

Figure 3.7. Coupler wavelength selectivity versus $\lambda(\mu m)$ (a) $\lambda_p = 1.248$ μm, $a = 2.55$ μm, (b) $\lambda_p = 1.422$ μm; $a = 2.71$ μm, (c) $\lambda_p = 1.636$ μm; $a = 2.55$ μm ($\lambda_p = \lambda$ peak) [After Parriaux et al. 1981b].

mental tuning curves at different wavelengths (Digonnet and Shaw 1982).

A second process for constructing a coupler makes use of fused biconical taper techniques, in which two or more fibres twisted around each other are heated and fused while the two stages of the fusion state are moved steadily apart (see Figure 3.9). The twist ensures that the fibres remain in contact. Such a coupler can be made with flame microburners or by microheaters controlled by electric current (Takeuchi and Noda 1992). The two fibres are secured to two movable plateforms. The tapering is stopped when the coupler has reached the desired transfer ratio.

Optical fibre coupling devices based on the fused biconical taper structure have been used as low loss branching points for multimode circuits (Agarwal 1985, Noda et al. 1987, Kawasaki et al. 1981). Many single-mode fibre couplers are produced by fused biconical taper

---

Figure 3.8. Experimental tuning curves of a coupler with corresponding theoretical fit: (a) $\lambda = 514.5$ μm, distance between cores $d = 5.24$ μm cladding index $n_2 = 1.4569$, $L < L_c$, (b) $\lambda = 632.8$ μm, $d = 5.4$ μm, $n_2 = 1.456$, $L = L_c$, (c) $\lambda = 632.8$ μm, $d = 4.77$ μm, $n_2 = 1.4578$ $L > L_c$ (overcoupling) [After Digonnet and Shaw 1982].

Directional couplers 85

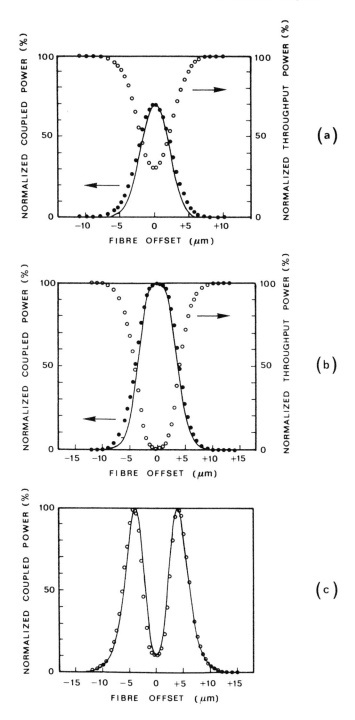

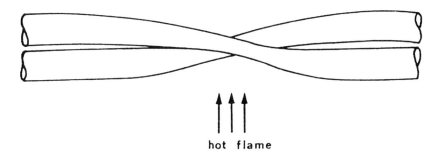

Figure 3.9. Fused biconical coupler.

structures (Georgiou and Boucouvalas 1985, Rawson and Nafarrate 1978, Saleh and Kogelnik 1988, Bures et al. 1984). This device exhibits low loss (0.5 dB), arbitrary branching ratio, polarization independence and broadband wavelength operation. Single-mode fused biconical couplers are also used for wavelength division multiplexing (Eisenmann and Weidel 1988).

In a fused biconical single mode fibre coupler, the coupling mechanism is not associated with evanescent waves and the infinite cladding approximation is not valid. A single guide is formed (Bures et al. 1983). The normalised optical power depends on the index n_3 of the external medium at various stages of fabrication, i.e. for various values of the waist radius ω (see Figure 3.10). An oscillating variation of the coupling factor as a function of the surrounding index n_3 is seen. The oscillation in power transfer becomes more rapid as ω decreases. The total power $P_1 + P_2$ remains constant except when n_3 approaches $n_2 (= n_3^{(c)})$. Variation of coupled power depends not only on surrounding refractive index but also on taper ratios (see Figure 3.11).

For evaluating the coupling ratio of fused biconical couplers the equation (3.8) in the approximation of weakly guided modes gives (for identical fibres) (Falciai et al. 1990):

$$C(z) = [NA\ u^2 K_0(w\ d\ /a)]/[n_1\ a\ V^3\ K_1^2(w)] \qquad (3.11)$$

In the coupler the propagation does not actually occur in the cores but in the common cladding of the two fused fibres, because in the coupler the fundamental mode in the input fibre is at cut-off ($V \ll 1$). The fibre coupling coefficient parameter C in equation (3.7) is replaced by those of the guide formed by the cladding and the surrounding medium (see Figure 3.12). Then we have:

$$C(z) = [(n_2^2 - n_3^2)^{1/2}\ u^2\ K_0(w\ d\ /b)]/[n_2\ a\ V'^3\ K_1^2(w)] \qquad (3.12)$$

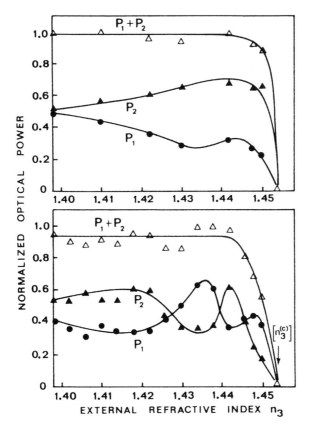

Figure 3.10. Characterization of a coupler as a function of the external refractive index for two values: (a) $2\omega = 8$ μm (waist diameter), (b) $2\omega = 4$ μm For $n_3 = n_3^{(c)}$ (cut-off value), P_1 and P_2 fall to zero [After Bures et al. 1984].

with

$$V' = (2\pi b/\lambda)(n_2^2 - n_3^2)^{1/2}$$

where b is the diameter and V' the normalized frequency of the new guide.

As V' is large, a simplified expression can be used (Ragdale and Goodman 1985):

$$C(z) = 1.018\, \lambda^{5/2}/\pi^3\, (n_2^2 - n_3^2)^{3/4} n_2\, [b(z)]^{7/2} \qquad (3.13)$$

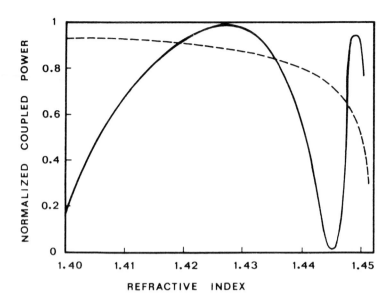

Figure 3.11. Variation of normalized coupled power P_i/P_{tot} versus refractive index: NA = 0.09, d/a = 20, P_i coupled power, P_{tot} total power.

$$\text{taper ratio } T_r = \frac{\text{initial fibre diameter}}{\text{diameter at the centre of taper}},$$

Broken curve — $T_r = 10$, solid curve — $T_r = 20$.

[After De Fornel et al. 1984].

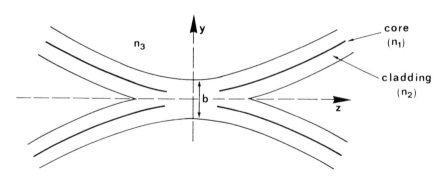

Figure 3.12. Schematic of fused coupler without cores.

The resulting coupling ratio is

$$CR = P_1/(P_1 + P_2) \qquad (3.14)$$

If L is the length of the coupler, the powers P_1 and P_2 are respectively proportional to

$$\cos^2 \int_0^L C(z)dz \text{ and } \sin^2 \int_0^L C(z)\,dz = CR \qquad (3.15)$$

Fused tapered single-mode 2×2 fibre couplers are available. They have a flat wavelength response and are thus useful in applications requiring a near-constant power-splitting ratio over a wide spectral range. The main difference between tapered fibres is their overall diameters.

A fused taper asymmetric multimode coupler consists of two dissimilar multimode fibres. The larger can be a bus fibre and the smaller a tap fibre. Only a small fraction of the bus power is coupled to a tap fibre, and so several similar fused tapers can be used along this bus fibre (Griffin et al. 1991).

Couplers made from unlike fibres may also be useful for integrating systems having different fibres, and can be used in sensors. For example, in some fibre lasers, pump light may need to be injected by a coupler at a wavelength that is well outside the single-mode range of the laser fibre (see Chapter 6).

The operation of asymmetric couplers depends not only on the level of asymmetry but also on the degree of fusion of the fibres. For two fibres with suitable initial diameters, the control over the degree of fusion of the coupler should be sufficient to produce the required power transfer without the need to etch or pretaper the fibres (Birks and Hussey 1988).

3.2.2 Y-coupler

The Y-coupler is a similar device to the X-coupler, and made by the same techniques: polished cladding, fibres brought together, fused coupler. The literature reports a number of designs for multimode and single-mode Y couplers (Belovolov et al. 1987, Kieli and Herczfeld 1986, Chattopadhyay and Nakajima 1990). For local area networks (LANs), asymmetric non-reciprocal fibre optic couplers are useful for construction of a linear bus. In order to cascade many such couplers, good power transfer between trunk ports 1 to 2 or 2 to 1 is necessary (see Figure 3.13).

Efficient coupling from the drop 3 to the trunk 1 or 2 is desirable in order to take full advantage of signals from a local transmitter. The

Figure 3.13. Scheme of a Y coupler.

coupling from the trunk to the drop should be small. The optical power between the three ports is given (in the case of a linear system) by:

$$\begin{bmatrix} P_{01} \\ P_{02} \\ P_{03} \end{bmatrix} = \begin{bmatrix} p_{11} & p_{12} & p_{13} \\ p_{21} & p_{22} & p_{23} \\ p_{31} & p_{32} & p_{33} \end{bmatrix} \begin{bmatrix} P_{i1} \\ P_{i2} \\ P_{i3} \end{bmatrix} \quad (3.16)$$

P_{0j} are the output powers and P_{ij} the input powers. The terms p_{13} and p_{31} are small, p_{21} and p_{12} are large and $p_{ii} = 0$.

When a Y junction is operating as a power combiner, at least 50% of the incident power in each leg is lost by radiation. An asymmetric non-reciprocal fibre optic coupler using multimode fibres with unequal core diameters showing a non-reciprocity $p_{23} > p_{32}$ due to modal-dependent coupling has been achieved (Kieli and Herczfeld 1986).

When a Y junction is used as a power splitter, a minimum insertion loss is 3dB, because 50 % of the incident power is coupled to one output port and 50 % to the other. In order to avoid excessive power losses caused by radiation, the angle between the two legs of Y must be very small (<1°). Low losses of 0,5-1 dB for splitting have been obtained with fused tapered fibres (Berolov *et al.* 1987).

3.2.3 Star coupler

A star coupler uses one mixing rod to distribute light evenly to a large number of terminals, provided all cable lengths are approximately equal. It can be designed to allow communication between every terminal of the system or to allow every terminal to communicate with a central processing unit. A single-mode fibre passive star coupler is an example. It can be used in architectures of high-speed optical local area networks (LANs). Many techniques are shown in the literature for constructing transmissive $N \times N$ star couplers (Rawson and Nafarrate 1978, Ozeki and Kawasaki 1976, Wilson and Bricheno 1990, Ohshima *et al.* 1985).

The evanescent field 3 dB coupler (2×2 coupler) previously mentioned can be used as a building block to construct a larger *n*-star, with *n* equal to an arbitrary power of 2 (see Figure 3.14). Reflective *N*-

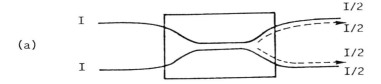

(a)

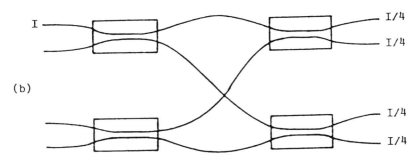

(b)

Figure 3.14. Star coupler (b) made from X couplers (a).

stars, where N is a perfect square, can also be constructed using 3dB couplers and mirrors, the number of components needed to realize a reflection star coupler is half that needed for an equivalent transmissive star coupler (see Figure 3.15). Examples are given by Saleh and Kogelnik (1988).

For an $N \times N$ star, in a network structure for interconnecting N terminals (i.e., with N input ports and N output ports), the number of crosspoints or switches is N^2. This number is a huge quantity for large values of N. Several one-sided optical switching networks minimizing the number of switching elements are described. When N is a power of 2, the number of stages is $\log_2 N$, and the number of 2×2 couplers required is $N/2 \log_2 N/2$ (Hill 1986). A reduction in the number of fibres can also be achieved using a technique in which the centralized transmissive $N \times N$ star coupler or the reflective N star coupler is replaced by a distributed version (Irshid and Kavehrad 1991).

In another fabrication process the fibres are bundled together. They are heated with an oxyhydrogen flame and pulled into a biconical taper shape which is cut at the waist. A cylindrical mixer rod is inserted between the tapers using the fusion-splice technique. The fibres are not twisted, and will be free from microscopic bendings which would lead to mode conversion. A 100×100 star coupler using standard

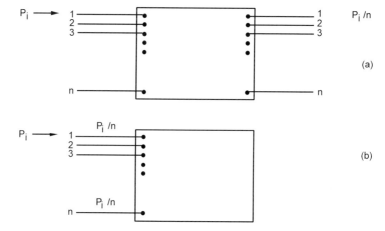

Figure 3.15. Scheme of star coupler: (a) transmissive, (b) reflective.

graded index silica glass multimode fibres with a waist diameter nearly 200 μm and excess loss 3.2 dB has been successfully fabricated (Ohshima *et al.* 1985).

A different approach uses radiative coupling between carefully configured bundles of tapered single mode fibres. The schematic arrangement of the two bundles is shown in Figure (3.16). In such tapered fibres, radiation is no longer confined to their cores, but fills the whole fibre cross section. For $1 \times N$ couplers, the optimum geometry is obtained when each fibre of the N arrays is directed towards the input fibre. In the case of $N \times N$ coupler, every fibre is directed towards the centre of the opposite array.

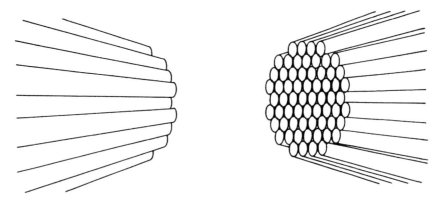

Figure 3.16. Schematic arrangement of angled 64×64 coupler [After Wilson and Bricheno 1990].

A new approach uses planar waveguides (Tabiani and Kavehrad 1991, Yanagawa *et al.* 1990) or two arrays of strip waveguides separated by a free-space planar waveguide propagating region (Dragone 1989, Dragone *et al.* 1989). With this technique, the losses are smaller and the loss benefit over a parallel geometry is 3dB for the $1 \times N$ star and 6 dB for the $N \times N$ star (Wilson and Bricheno 1990).

Switching networks with star coupler can be used in order to interconnect a large number of optical terminals (Hill 1986). Star couplers can also be used with active fibre amplifiers (Irshid and Kavehrad 1992, Willner *et al.* 1991, Liaw *et al.* 1993). The advantages of the star coupler have already been pointed out, but for future developments it will be necessary to reduce costs and improve performance (Lu *et al.* 1990).

3.2.4 Micro-optics coupler

Another category of optical fibre couplers is based on micro-optic components: cylindrical GRIN rod lenses, spherical retro-reflecting mirrors using two fused silica microprisms suitable for multimode fibres with a large core, or two optical fibres with end faces cut at an angle of 45°. (see Figure 3.17b, c) (Kuwahara *et al.* 1975, Tomlinson 1980, Suzuki and Kashiwagi 1976).

3.3 BRAGG GRATINGS

3.3.1 Production

Gratings are useful components for many devices both passive and active. In these ones, the grating can be either an external component or written directly into the fibre.

To produce a fibre grating resonator, a single-mode fibre is lapped and polished to gain access to the field in the core (for details see § 3.3.2). It is then coated with a thin layer of photoresist. A moiré grating is formed in the photoresist by successive exposures to two interference patterns of slightly different periods. After developing the photoresist, the grating is etched, coated with a thin layer of aluminium oxide and covered with index-matching oil. The resonator has a linewidth of 0.04 nm near 1500 nm (Reid *et al.* 1990).

Bragg gratings can be also realised in germanium–silicate fibres by a transverse holographic method, i.e., by a single exposure of the core from the side of the fibre to a two-beam interference pattern. To write these gratings the sheath material is temporarily removed for exposure. The source is a coherent UV beam from a pulsed laser tuned

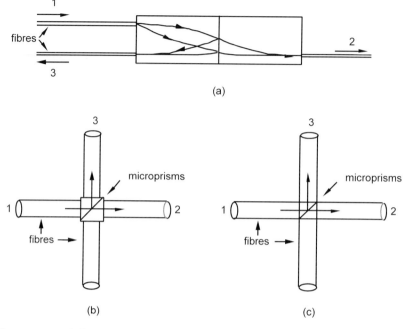

Figure 3.17. Micro-optic coupler: (a) GRIN lens, (b) Microprisms, (c) fibre end face cut at 45° angle.

to a wavelength lying in the 244 nm germanium–oxygen vacancy defect band. The grating period Λ is fixed by the source wavelength and the angle θ of the interfering beams by $\Lambda = \lambda_L/(2\sin(\theta/2))$. Reflectances from 55%–94% can be obtained for several wavelengths (570 nm–1500 nm) using different lengths (Meltz et al. 1989, Ball et al. 1990, Kashyap et al. 1990). Grating reflectances as high as 99.5% are possible. The highest grating length, 15 mm to date, gives narrowest spectral bandwidth. For example, in high germanium-doped fibres, index modulation changes can reach 1.8×10^{-3} and depending on the intensity of the exposing beam and the exposure duration. The grating forms a band-blocking filter, as shown in Fig. 3.18, by reflecting optical signals of wavelength λ_L in the light guiding core according to the Bragg condition: $\lambda_L = 2n_e\Lambda$ where Λ is the grating spacing and n_e the refractive index defined by equation (1.25).

The grating spacing and thus the Bragg wavelength depends on the angle of each of the exposing beams on the side of the fibre, and on the exposing beam wavelength. Reflection gratings have been fabricated with Bragg wavelengths from the visible spectrum to the 1550 nm IR communication bands. These gratings can be erased in the

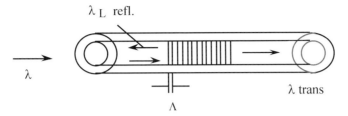

Figure 3.18. Illustration showing fibre grating reflector and Bragg condition for reflection.

same way by another photorefractive effect (Niay et al. 1994). Gratings can also be written in elliptical-core HiBi (High Birefringent) fibres by the transverse holographic method (Niay et al. 1995).

In a broadly similar way, the feasibility of writing permanent moiré photorefractive gratings directly into the core of a germanosilicate fibre has been demonstrated by using two successive transverse illuminations of the fibre by UV fringe patterns of slightly different periods. The grating filter consists of two intra-core Bragg reflection gratings separated by an optical phase shift (Fertein et al. 1991, Legoubin et al. 1991).

Absence of germanium reduces photosensitivity in optical fibres, so Bragg gratings are easier to write in germanium-doped fibres. Photosensitivity could be improved by loading of fibres with molecular hydrogen. This process depends on both temperature and pressure. The best results for Bragg grating writing, i.e., with highest index change, have been obtained with germanium-doped fibres in cold hydrogen under high pressure (Kashyap 1999). Writing gratings in rare-earth doped fibres is more difficult than in standard fibres because they do not contain any germanium.

High-performance long-period fibre gratings, based on induced periodic microbends can be made using an electric arc (Hwang et al. 1999). Bragg gratings can also be written in multimode fibres (Mizunami et al. 2000).

The applications of all-fibre components are various: wavelength multiplexing and demultiplexing (see § 3.4), frequency shifting, wavelength filtering, dispersion compensation (Eriksson et al. 1994, Williams et al. 1995), sensors (Morey et al. 1992, Brady et al. 1994) fibre lasers (see § 6.2), compressors (Karlsson 1994) and as part of delay lines (Ball et al. 1994). Multiple mode conversion LP_{01} and LP_{02} can be induced by periodic coupling through a fibre grating and through a Fabry-Perot mirror (Shi 1992) and a wide band Fabry-Pérot-like resonator may be fabricated using two chirped Bragg gratings (Town et al. 1995).

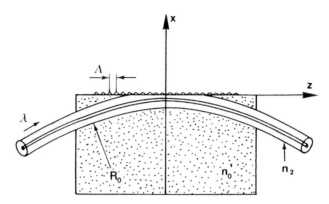

Figure 3.19. Scheme of a grating fibre coupler. The y axis is perpendicular to the plane of the drawing [After Russel and Ulrich 1985].

A grating–fibre coupler consists of a fine relief grating of period Λ formed usually in photoresist on the surface of a side-polished single-mode fibre (Russel and Ulrich 1985). The fibre is cemented on to a curved groove in a glass block. The block and the fibre are polished, removing part of the cladding, in order to have access to the evanescent field of the fibre (see Figure 3.19). Experiments are using gratings, holographically defined in deposited photoresist layers or metal gratings not stamped on to the polished fibres.

3.3.2 Theory

The different beams travelling into the fibre and coupled outside have wave vectors with z components. If $K = 2\pi/\Lambda$ is the grating constant and Λ the spatial period, the interaction of light with the grating direction α_m is given using the components of the wave vectors with z (see Figure 3.20b).

$$(\mathbf{k_m} - \mathbf{k_e}) \cdot \Lambda = m\, 2\pi \qquad (3.17)$$

with

$$\Lambda = \Lambda\, \mathbf{e_z} \qquad |\mathbf{k_m}| = (2\pi/\lambda_0)n_i \qquad |\mathbf{k_e}| = (2\pi/\lambda_0)n_e \qquad (3.18)$$

where $n_e = \beta/k_0$ is the mode effective index, n_i the superstrate index ($n_i = 1$ for air) or the substrate index ($n_i = n_0$, the silica glass index); λ_0 is the wavelength in empty space; m denotes the diffraction order.

Equation (3.17) gives:

$$n_i \sin \alpha_m - n_e \sin \alpha_1 = m\, \lambda_0/\Lambda \qquad (3.19)$$

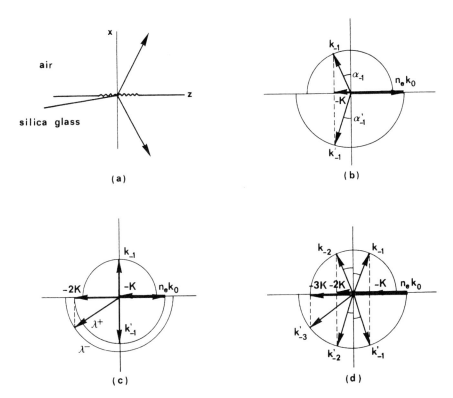

Figure 3.20. Grating fibre coupler: (a) direction of beams, (b) $\lambda_0 > \lambda_c = n_e \Lambda$; only $m = -1$ is possible, a light directed backward; (c) $\lambda_0 = \lambda_c$. Light is in plane xy, (d) $\lambda_0 < \lambda_c$ several orders can propagate in both forward and backward directions.

The angle α_1 equals approximately $\pi/2$ because the direction of modes propagating in the fibre are nearly perpendicular to the x-axis and in the case where the substrate is air, we have:

$$\sin \alpha_m = n_e + m\lambda_0/\Lambda \tag{3.20}$$

and in the glass ($n_i = n_1$)

$$\sin \alpha'_m = n_e/n_1 + m\lambda_0/\Lambda n_1$$

so light is diffracted.

When $\lambda_0 > n_e \Lambda$ output radiation can leave the coupler only in two half cones of order $m = -1$, with k_{-1} directed backward.

If $\lambda_0 = n_e \Lambda$, these half cones become the plane $\alpha_{-1} = 0$, which is the xy plane of Figures 3.19 and 3.20a. Power in the $m = -2$ order is reflected into the input fibre. When $\lambda_0 < n_e \Lambda$ several output orders can propagate in both forward and backward directions.

Russel and Ulrich (1985) have described a spectrometer based on a grating fibre coupler. A photoresist grating is placed in the evanescent field near the core. When light from a tunable laser is injected into the fibre two fan-shaped output beams are observed. The angle of deviation, in air ($n_1 = 1$), varies with λ_0 following the equation

$$\sin \alpha_{-1} = n_e[1 - \lambda_0 / (n_e \Lambda)] \tag{3.21}$$

For example, with $n_e = 1.459$, $n_1 = 1.55$, $\Lambda = 434$ nm, results are shown on table 3.1.

Table 3.1 Values of α and α' for different orders m

	$\lambda = 1$ μm		$\lambda = 0.6328$ μm		$\lambda = 0.4$ μm	
m	$\alpha(°)$	$\alpha'(°)$	$\alpha(°)$	$\alpha'(°)$	$\alpha(°)$	$\alpha'(°)$
1	—	—	—	—	—	—
0	—	—	—	+70.17	—	+70.27
−1	−57.69	−33.04	0	0	+32.50	+20.28
−2	—	—	—	−70.07	−22.60	−14.36
−3	—	—	—	—	—	−57.41

Rowe et al. 1987 have studied the effects of surface relief gratings in single-mode fibres. Figure 3.20c also illustrates the effect of wavelength variations around those fitting the Bragg condition. The grating vector providing phase matching at the first order Bragg condition is defined by $K = 2\beta$ (or $\lambda = 2n_e\Lambda$). In this representation, the semicircle radii and β vary inversely with λ. At wavelengths λ longer than that specified by the equation $\lambda = 2n_e \Lambda$, the tip of the K-vector ($K = 2\pi/\Lambda$) lies beyond the tip of the β vector $K > 2\beta$, no phase-matching condition is possible and the grating exerts no influence on the propagating mode. By moving to shorter wavelengths however, the condition $K < 2\beta$ is achieved and the grating induces radiation into the cladding. The range of wavelength between the peak of the Bragg condition and the onset of radiation is clearly determined by the size of $n_e - n_2$.

Structures employing high refractive index overlayers (Al_2O_3/oil) are discussed. Reflectances up to 98% and linewidths of 0.8 nm at the first order have been achieved. Figure 3.21 shows the measured response as a function of wavelength. Applications of these Bragg gratings could be as different as multiplexing (§ 3.4), filtering (§ 3.6), resonating (§ 3.9.1) or different laser reflector (Chapter 6).

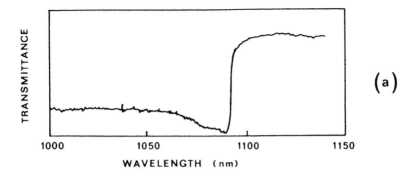

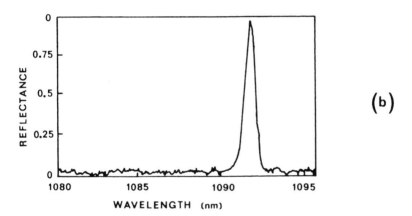

Figure 3.21. Measured response of a fibre grating as a function of wavelength: (a) transmission, (b) reflection [After Rowe *et al.* 1987].

3.4 WAVELENGTH MULTIPLEXERS AND DEMULTIPLEXERS

3.4.1 With external components

More efficient utilization of the large transmission capacity of optical waveguides is obtained by the wavelength division multiplex mode in fibre optics communications systems and sensors. Experimental fibre optics links operating in the wavelength division multiplex mode have already been used successfully (Mahlein 1983).

The basic wavelength division multiplex (WDM) concept is the simultaneous transmission of the modulated emission of several light sources operating with different wavelengths over a single optical

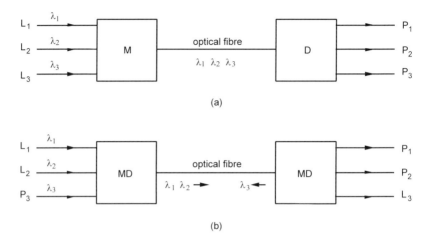

Figure 3.22. Scheme of multiplexer and demultiplexer: (a) unidirectional lines, (b) bidirectional lines.

waveguide. An example of an unidirectional communication link operating with three wavelength channels is shown in Figure 3.22a. For the combination and separation of the light of the various wavelength channels at the beginning and end of the link an optical multiplexer (*M*) and demultiplexer (*D*) are used. For example, three diodes (LEDs, or laser diodes) L_1, L_2, L_3 send the light into the multiplexer and then into the fibre. Three PIN or avalanche diodes receive the signal through the demultiplexer MD (see Figure 3.22b). It acts as a multiplexer for some of the channels and as a demultiplexer for some of the others. The numbers of 'Go' and 'Return' channels need not be equal.

Various physical principles are available for the practical realization of the MD devices: image formation by lenses or concave mirrors, beamsplitting principles (Winzer et al. 1981), material dispersion by interference with multilayer structures, or diffraction by gratings (see Figure 3.23). In this last case, for example, a length of step-index single-mode fibre is cemented in a convex groove in a fused silica block. Following polishing and cleaning a surface relief grating with a period Λ is added (Rowe et al. 1987, Russel and Ulrich 1985, Jauncey et al. 1986).

Another method for wavelength division multiplexing has been reported using holographic gratings produced in deposited photoresist layers, or metal gratings in contact with polished fibres. Reflectivities of 98% at the first order Bragg wavelength and linewidths of 0.8 μm are achieved. These applications are also in laser mode selection, switching

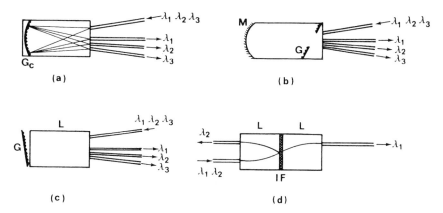

Figure 3.23. Several examples of WDM couplers: (a) with concave grating G_c, (b) with plane grating G and mirror M, (c) with plane grating G and graded index lens L, (d) with an interference filter IF and two graded index lenses L.

and nonlinear fibre modulators. The degree of suitability for a given application depends on the physical principle of the devices. They require low insertion loss, compact design and high reliability (Whalen and Walker 1985).

In optical WDM (wavelength division multiplexing) systems, several independent digital signals are transmitted in a single fibre at different wavelengths. Each signal can be affected by noise or by cross-talk from neighbouring channels (Geckeler 1990, Loeb and Stilwell 1990, Rocks 1987). A polarization-independent narrow channel WDM fibre coupler for operation in the 1.55 μm wavelength region has been developed (Mc Landrich et al. 1991). Dense WDM for systems is achieved by devices where the wavelength spacing is of the order of 1 nm; optical frequency division multiplexing (FDM) by systems where the optical frequency spacings are of the order of the signal bandwidth or bit rate. A review of WDM systems is given by Brackett (1990).

3.4.2 All-optical-fibre devices

3.4.2.1 Using Bragg grating devices

In the first method the channel signal is split into several branches by a star coupler and in each branch end a Bragg grating selecting a specific wavelength is written (Agrawal and Radic 1994) (see Figure 3.24). Fibre gratings are currently proposed for WDM (Liaw et al. 1999; see also § 3.6).

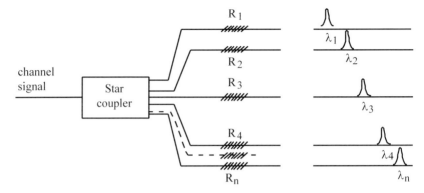

Figure 3.24. Scheme of an all fibre WDM.

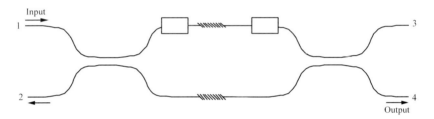

Figure 3.25. Bragg gratings written into a Mach-Zehnder interferometer.

Another method uses a Mach-Zehnder interferometer with Bragg gratings written into the fibres constituting the arms of this one. Coupling efficiency of 99.4%, transmission loss <0.5 dB, return loss of 23 dB have been achieved (Bilodeau et al. 1995) (see Figure 3.25).

3.4.2.2 Using mechanical devices

An in-line fibre tap has been fabricated from bimodal fibre supporting light propagation for two modes (LP_{01} and LP_{11}) at the operating wavelength. This tap consists of two fibre devices: a mode converter, and a fused directional coupler (see Figure 3.26). Light entering the tap propagates in LP_{01} mode. Passing through the mode converter a part of this light is converted into LP_{11} mode. The periodic perturbation may be stress-induced, step or holographically written. The mode converter is designed to induce mode conversion for only a narrow band of wavelength. The fused coupler is designed to couple the LP_{11} mode of the first fibre into the mode of the second optical fibre, whereas light in the LP_{01} mode is uncoupled and travels through the coupler with low

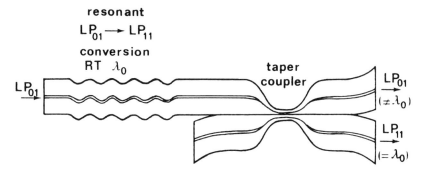

Figure 3.26. Bimodal fibre narrow-band tap [After Hill et al. 1990].

loss. The mode converter is fabricated by microbending, inducing coupling between LP_{11} and LP_{01}.

A Mach-Zehnder interferometer constructed with a twin-core fibre can also be used for demultiplexing two wavelengths. The coupling power for one of the two wavelengths is achieved by bending the fibre in the plane of the twin cores (Arkwright et al 1993).

3.5 FREQUENCY AND PHASE SHIFTERS

3.5.1 Frequency shifters

In an unperturbed straight fibre the two modes LP_{01} and LP_{11} are orthogonal, and do not exchange power as they propagate along the fibre. In interferometric and sensor applications, periodic coupling between the modes in an optical fibre is important; it gives rise to frequency-shifting effects. Coupling between the LP_{01} and LP_{11} modes can be achieved by introducing microbends into the fibre, by periodically squeezing it, or by an acoustic wave travelling inside the fibre (see Figure 3.27) (Birks et al. 1996).

An all-fibre optics frequency shifter can be made using mode coupling between LP_{01} and LP_{11} modes by a travelling acoustic flexural wave, guided along the optical fibre (Kim et al. 1986) (see Figure 3.28). Coupling between modes is achieved when the phase matching condition $\Lambda = L_b$ is fullfilled.

Λ is the acoustic wavelength and L_b the fibre beat length ($L_b = \lambda_0/\Delta n$, where λ_0 is the optical wavelength and Δn is the refractive index difference between the two polarized modes). In this case, the optical signal coupled from one mode to the other is shifted in frequency (Risk et al. 1984, Heismann and Ulrich 1984, Yijiang 1989). The signal coupled from the slow mode LP_{01} to the fast mode

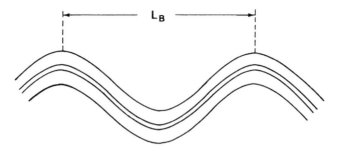

Figure 3.27. Modal coupler made with squeezed fibre.

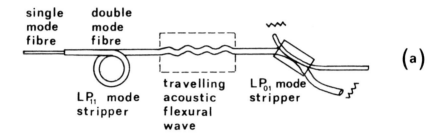

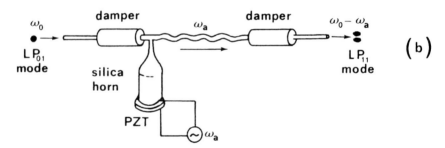

Figure 3.28. (a) Schematic diagram of an all-fibre-optics frequency shifter with mode filter for the LP_{01} and LP_{11} modes and a travelling acoustic flexural wave, (b) Frequency shifting in a double-mode fibre using intermodal coupling by an acoustic flexural wave excited by an acoustic horn [After Kim et al. 1986].

LP_{11} is shifted down in frequency when the acoustic wave is propagated in the same direction as the optical signal, and shifted up if the acoustic wave is travelling in the opposite direction to the signal one, for coupling from the fast mode to the slow mode.

In another device an acoustic wave with an acoustic wavelength Λ_s ($\Lambda_s = v_s / f_s$ where v_s and f_s are respectively the acoustic phase velocity and frequency) is launched along a fibre containing a strong Bragg grating. The overlap of acoustical and optical power can be increased by reducing the fibre diameter. In this device the forward and backward propagating modes in the gratings are coupled and operate in counter-directional reflection modes. A guided mode incident on a Bragg grating carrying a counter-propagating acoustic wave is reflected into a counter-propagating mode with a Doppler shift $\pm fs$. This device is a frequency shifter in reflection and it can also be a tunable filter or a switch (Liu et al. 1998).

A birefringent fibre guides two orthogonally polarized modes, and owing to the large difference between their propagation constants, there is little coupling between these modes. However, if the fibre is periodically stressed, complete transfer of power from one polarization to the other can occur. Passive and active couplers have been demonstrated (Youngquist et al. 1985, Risk et al. 1986). By using a travelling acoustic wave to produce a spatially periodic stress in the fibre, light can be coupled between the two principal polarizations of the birefringent fibre. The phase matching condition is the same as above ($\Lambda = L_b = \lambda_o / \Delta n$). The power coupling from one polarization to the other reaches a peak when the optical wavelength is such that the beat length of the fibre matches the acoustic wavelength. By changing the acoustic frequency, the centre wavelength of the optical passband can be tuned. This device can also act as wavelength filter. It is described under WDMs (see § 3.4).

Another LP_{01} to LP_{11} modal coupler uses periodic microbends spaced by a beat length defined as $L_b = 2\pi/\Delta\beta$, where $\Delta\beta$ is the difference of the propagation constants between the two modes LP_{01} and LP_{11} along the fibre. For example, the beat lengths are 270 mm and 265 mm at $\lambda = 590$ nm and $\lambda = 496.5$ nm when the fibre used has a core radius of 2.28 μm and a cut-off wavelength for the second mode of $\lambda_c = 671$ nm. The maximum coupling to LP_{11} mode achieved so far has been 99.68 % (Blake et al. 1986, Blake et al. 1987).

Wavelength tuning can be achieved by applying a small, controllable flexure to a long period fibre grating (Van Viggeren et al. 2001) or by pressing a plate with periodic grooves against a short fibre length (Savin et al. 2000).

3.5.2 Phase shifters

A thermo-optic phase shifter can be constructed by coating the arms of the interferometer with resistive ink. On the application of a voltage between the ends of the coated region, the temperature of the fibre arm of length L is raised owing to resistive heating. The differential phase shift is:

$$\frac{1}{L}\frac{d\phi}{dT} = k\left(\frac{n_e}{L}\frac{dL}{dT} + \frac{dn_e}{dT}\right) \qquad (3.22)$$

A second phase shifter can be created by longitudinally stressing one fibre arm. A strain εL induced by a force F along the fibre length L is given by $\varepsilon L = F/AE$, where E is the Young's modulus for silica and A is the cross section area. The optical phase shift is:

$$\Delta\phi = \frac{\beta F L}{AE}\left[1 - \frac{n_e^2}{2}\{(1-\sigma)p_{12} - \sigma p_{11}\}\right] \qquad (3.23)$$

where σ is Poisson's ratio and p_{11}, p_{12} the Pockels coefficients. Longitudinal stress is obtained using a flexing translation. Two piezoelectric ceramic plates act in opposition, causing a shear stress on the fibre.

Phase shifting can be also created using fibre Bragg gratings (Agraval and Radic 1994).

3.6 WAVELENGTH AND MODAL FILTERS

3.6.1 Wavelength filters

As seen at the beginning of this chapter, recent theoretical and experimental investigations have led to wavelength filtering devices for WDM operation and for light source spectral filtering. Optical edge filters transmit light for wavelengths $\lambda < \lambda_0$ or $\lambda > \lambda_0$, λ_0 being the specified band-edge. They are usually designed as interference filters. Different edge filters made of single-mode fibres have been fabricated. In § 3.2.1 of this chapter concerning the construction of X couplers, we have seen that these devices show a power division function of wavelength λ. The fundamental mode propagates under ideal conditions losslessly with a phase constant β and an effective index $n_e = \beta/k$ ($k = 2\pi/\lambda$). Both β and consequently n_e depend on the wavelength λ. If part of the cladding has a refractive index n_3 greater than n_e, it acts as a mode sink, and power leakage occurs from the guided light. This effect depends on the wavelength. The power leakage increases as the wavelength increases because the field of the mode extends into the cladding and the difference $n_3 - n_e$ increases (see Figure 3.29). Insertion losses are below 1 dB in the transmission bands and attenuations are better than -30 dB in the stop bands.

In another device the waveguide is overclad by a silica superstrate n_4 equal in index to that of the fibre cladding n_2 (see Figure 3.30). For efficient directional coupling, phase velocity matching

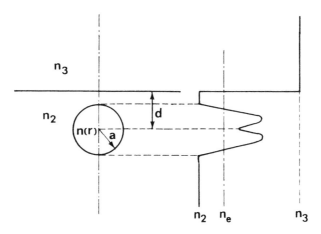

Figure 3.29. Transverse cross-section of a single-mode fibre with a mode sink [After Zengerle et Leminger 1985].

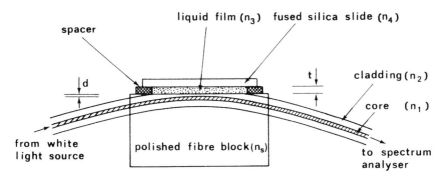

Figure 3.30. Experimental construction of a channel-dropping filter [After Millar *et al.* 1987].

conditions require that the film and the fibre mode refractive indices be equal ($n_{3e} = n_e$). Guidance in the waveguide and wave coupling occurs when $n_3 > n_{3e} = n_e > n_4 = n_2$.

The wavelength λ_m at which the fibre mode is phase-matched to the forward film mode is given by:

$$\lambda_m \cong \frac{2d}{m}\left(n_3^2 - n_e^2\right)^{1/2} \quad (3.24)$$

where m is the order of the mth propagating mode at λ_m in the direction of the fibre axis in the symmetric planar waveguide of thickness d. If the interaction length of the fibre–film coupler is equal to the coupling

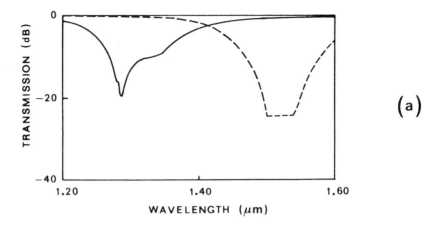

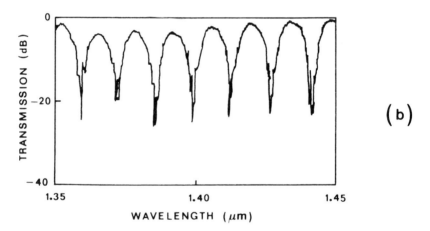

Figure 3.31. (a) Response for a thin oil overlay waveguide (thickness 0.91μm), (b) channel-dropping response for a thick overlay waveguide (thickness 81 μm) [After Millar et al. 1987].

length, light at wavelength λ_m couples out of the fibre into the overlay guide. With a thin overlay waveguide the device has a single tunable dropped band response (see Figure 3.31a) and with a thick overlay waveguide it has a comb filter response with a channel spacing of 13 nm (see Figure 3.31b) (Miller et al. 1987).

Another technique for creating a wavelength filter is based on the use of coaxial couplers–dissimilar waveguide couplers. The coupled waveguides are a rod and a tube and can thus be represented by the

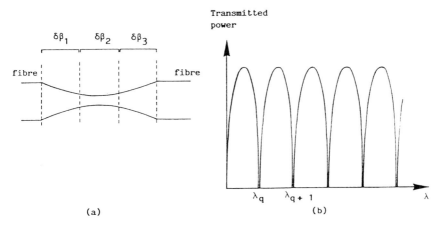

Figure 3.32. (a) Tapered coaxial coupler, (b) transmitted power of the filter [After Boucouvalas and Georgiou 1987].

core and cladding of a fibre. Significant power transfer between the core and cladding of a single-mode optical fibre has been demonstrated by the tapering technique. A tapered coaxial coupler represented in Figure 3.32a appears as alternating $\delta\beta$ couplers, if the taper thickness is below a value defined by the phase matching thickness of the coupled waveguide (Boucouvalas and Georgiou 1987). When the coupling begins, the transmitted power oscillations appear and after a desired number of oscillations the process of tapering is stopped. This number of oscillations determines the period of the coupler used as a filter. A (rectified) sinusoidal curve fits the taper wavelength response λ_m (Figure 3.32b).

The transmission $P(\lambda)$ can be written as:

$$P(\lambda) = \frac{1}{2}\left[1 + \sin\left\{\frac{2\pi}{\delta\lambda}(\lambda - \lambda_m)\right\}\right] \quad (3.25)$$

where $\delta\lambda$ is the period of the tapered filter response.

In a similar way, when a concentric ring with an index higher than the cladding one is added to a conventional core, the structure supports an additional symmetric mode with substantial power in the ring. The outer ring and the central core play the respective parts of the two cores in a mismatched fibre filter core (Dong et al. 2000).

Another kind of selective fused coupler consists of a mismatched twin core fibre and a standard fibre. This device acts as a filter or as a switch (Ortega and Dong 1999).

A tunable filter system is based on two piezoelectric stack actuators moving a mechanical device allowing both traction and compression thus setting up an apodized fibre Bragg grating. The device can work in transmission and in reflection (Iocco et al. 1999).

Fibre-optics filters based on coupling between two dissimilar fibre waveguides have also been reported (Marcuse 1985). A dissimilar two-core fibre segment can be used as a channel dropping filter based on the principle of grating-frustrated coupling. The grating can be located either outside or within the coupling region (Jacob-Poulin et al. 2000).

A spectral filter directly integrated into a single-mode fibre is made by cascading biconical fibre tapers, converting a single-mode optical fibre into a wavelength filter system (Lacroix et al. 1986, Wang 1987).

Low-loss structures employing gratings etched into polished fibre and overlaid with high refractive index coatings (see Figure 3.29) provide more than 90% reflectivity into the first Bragg order at 1.3 μm and can serve as a filter, mirror, non-linear modulator or switch (Rowe et al. 1987, Bennion et al. 1986, Ragdale et al. 1990).

3.6.2 Filters based on modal filtering

A device uses phase-matched evanescent coupling to transfer energy between a single-mode fibre and the LP_{11} mode of a two-mode fibre (Sorin et al. 1986). The transfer of power to the unwanted mode is minimized by making use of both the phase mismatch and the smaller modal overlap between the two lowest modes (Morishita 1991). Such a device is an in-line all-fibre modal filter with greater than 90% coupling to the LP_{11} mode. Suppression of the coupled power to the lower order LP_{01} is better than -24 dB. A similar device using two elliptical core fibres has been suggested by Kumar et al. (1990).

Other techniques for creating a filter require either a fibre in which guided acoustical and optical waves can propagate simultaneously by means of a piezoelectric transducer bonded to the fibre (Jen and Goto 1989) or a two-mode fibre around which helical microbending is produced by winding a length of wire along it. This last system induces strains in the same way as a grating grooved on the fibre (Poole et al. 1991).

3.7 LINEAR SWITCHES AND TAPS

3.7.1 Switches

Optical switches with low losses, small size and low driving power are required in transmission systems and signal processing. A key functional element for an advanced single-mode optical fibre network is a 2×2 cross-point switch for routing optical signals. An optical waveguide switch can be a device to turn the power on and off in a waveguide. This function is most simply performed electrically at the light source itself. A more general definition of an optical waveguide switch is a device for diverting the power from one optical fibre to any other in a number of adjacent waveguides.

An optical signal may be switched from one optical channel to another in three basic ways. In the first method a mirror, prism or other mechanical device may be placed in the path of the beam to be deflected, e.g., in the beam between two fibres (Young and Curtis 1981). Another possibility is a micro-electromechanical system (MEMS) technique using hinged flapping micro-mirrors in a linear array, and piston type micro-mirrors in a two-dimensional array (Riza et al. 1999). Figure 3.33 is a schematic diagram of an electrostatically driven fibre-optic micro-mechanical on/off switch. The elements are a membrane and a metal substrate, an insulator and input and output fibres. The membrane tip is inserted into a small gap between the two fibres. Light is transmitted between the fibres through a small hole in the membrane. When a voltage is applied between the membrane and the metal substrate, the membrane is distorted by electrostatic force. The life of a switch of this type can be as much as 10^5 cycles (Hogari and Matsumoto 1990).

It is also possible to construct an optical switch based on the *electrowetting* phenomenon. This is a surface tension effect caused by shear forces tangential to the interface (and is not the same as capillarity, which is caused by a force perpendicular to the interface). A capillary tube containing an electrolyte and mercury is sandwiched between two quarter-pitch GRIN lenses. By the electrowetting effect, a voltage applied between two electrodes at the ends of the tube causes a displacement of the mercury in or out of the light paths (see Figure 3.34). This switch has been tested for more than 10^7 cycles with no degradation in performance (Jackel et al. 1983).

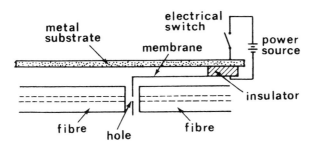

Figure 3.33. Electrostatically driven fibre-optic micro-mechanical on/off switch structure [After Hogari and Matsumoto 1990].

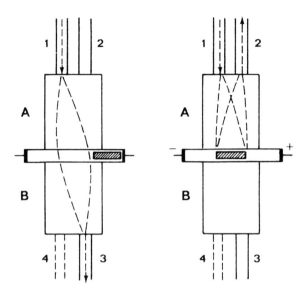

Figure 3.34. Fibre optical switch using continuous electrowetting [After Jackel *et al.* 1983].

The presence of moving parts implies slow switching, so these devices are used in sensors rather than in communications systems: The second method of switching gives a quicker response. The optical signal is converted into an electrical signal by a photodetector; this electrical signal is then switched.

The third method of optical waveguide switching utilises deflection of the optical power itself. In an optical cross-point switch using a high-index waveguide sandwiched between two polished fibre coupler blocks, the interaction length varies from 500 μm to 4 mm. The refractive index of the interlay waveguide is higher than the effective mode index of the fibres. Switching is effected by the input radiation coupling through the interlay waveguide into the second fibre (cross-coupled state) or recoupling into the first fibre (straight-through state). The switching action is obtained by varying the waveguide thickness (e.g., using a piezoelectric tranducer) or by varying the waveguide refractive index using a liquid crystal interlayer (see Figure 3.35).

The cladding of the fibre is polished down to within about 10–20 μm of the core. This device shows a spectrally varied response with peaks in the transmitted power at several wavelengths. The behaviour of the device at the transmission peak is dominated by the beating of two modes of opposite L_c symmetry (as in the fused tapered coupler). In the case of a liquid crystal interlayer transducer, the relationship between crystal

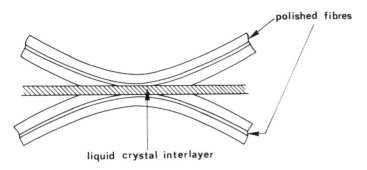

Figure 3.35. Interlay coupler.

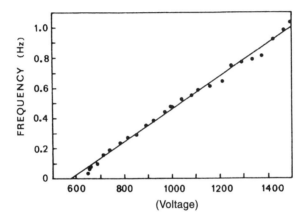

Figure 3.36. Plot of frequency versus voltage for the crystal pulsations • experimental data, — theoretical fit [After Goldburt and Russel 1986].

pulsations frequency and voltage is approximatively linear (see Figure 3.36) (Goldburt and Russel 1986, Wright et al. 1988).

Coupling of unpolarized light from one optical fibre to one of two other fibres can be achieved using a 2×2 optical switch with chiral liquid crystals and switchable waveplates (Shankar et al. 1990). Another method consists of using two 'D' fibres (polished to near the core) laid at a small angle to one another, their flat surfaces in contact (see Figure 3.37). The switching is achieved by small movements of this contact. Coupling efficiency of the switch is dependent on the angle between the two fibres (Cassidy and Yennadhiou 1988).

Shipley et al. (1987) have described a single-mode optical fibre switch that is electrically activated. This device is a Mach-Zehnder interferometer (see § 3.10), made with two couplers in series with a pair of single-mode fibres. It differs from the standard switch in that

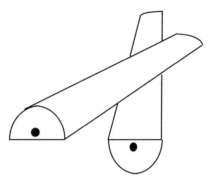

Figure 3.37. Scheme of 2D fibre switch.

either one or both of the couplers are constructed to have a beamsplitting ratio that is dependent on wavelength. Power splitting in the output arms is achieved by introducing a phase shift or path difference by the thermo-optic effect or by a piezoelectric ceramic that produces a flexing output.

When an acoustic wave is applied laterally to a fibre grating, it induces lateral vibrations and hence microbending of the fibre grating. The microbending serves as a long period grating for coupling core and cladding modes, so the reflection window of the fibre grating can be switched between the Bragg wavelength and the cladding mode coupling wavelength (Liu *et al.* 2000).

3.7.2 Taps

In-line optical fibre taps have been demonstrated using angled fibre mirrors placed into fibres (Shin *et al.* 1989), using bimodal fibre (Hill *et al.* 1990), or acousto-optic mode coupling (Patterson *et al.* 1990, Heffner and Kino 1987). An acousto-optic tap permits the amount of light taken from the fibre to be varied electronically by use of a phase array with acoustic transducers fabricated directly upon the surface of a standard cylindrical fibre. The Bragg condition allowing light to be deflected out the fibre by the acousto-optic interaction is $\sin \alpha = k_a/2k$ where k_a is the acoustic wave propagation constant, α the angle between the acoustic wave vector $\mathbf{k_a}$ and the optical propagation vector $\mathbf{k}$. The longitudinal waves are excited by the transducer phase array of acoustic transducer (with a period d) at angle $\pm \gamma$ to the normal, given by $\sin \gamma = 2\pi \, d/k_a$. If f is the centred frequency and V_a the acoustic velocity of the acoustic wave, by equating α and γ we obtain $f = V_a(kd/\pi)^{1/2}$ (Heffner and Kino 1987).

Acousto-optic mode coupling from the fundamental mode to a higher order core or cladding mode has been achieved using two optical fibre switchable taps (Patterson et al. 1990). The coupling is maximum when the interaction length L is equal to the beat length $L_b (= 2\pi/\Delta\beta$, where $\Delta\beta$ is the difference in mode propagation constants arising from intermodal dispersion). The coupling will tend to be shifted in phase as the interaction length becomes much greater than the beat length, and the coupling efficiency will decrease. A high frequency acoustic wave is introduced into the core at a slight angle θ measured from the plane normal to the fibre axis. The acoustic frequency is selected so that its propagation constant $k\theta$ in the z–direction (fibre axis) is equal to the difference in the mode propagation constants $\Delta\beta$ of the optical beat wave.

3.8 MODULATORS

3.8.1 Phase modulators

A modulator acts on the phase of the lightwave. When a strain is produced in a single-mode fibre, the phase of the light changes (De Paula and Moore 1984). A simple modulator is made of a sheathed optical fibre wound on a hollow cylindrical piezoelectric transducer (PZT) (Wysocki et al. 1989) and is used in interferometers to correct phase drift. All-fibre modulators using polymer or ceramic and optical fibres have been developed (Martini 1987). One method involves bonding a fibre into a piezoelectric plastic film. Strain is induced in the plastic film by an applied electric field and transmitted to the fibre. Fibres coated with a polyvinylidene fluoride (PVDF) or vinylidene fluoride copolymer sheath have been developed. The sheathed fibre and coaxial electrodes forms of a long, thin layered cylinder (Imai et al. 1988, Gusarov et al. 1996, Roeksabutrand Chu 1996) (see Figure 3.38).

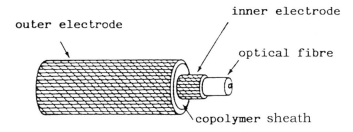

Figure 3.38. Schematic diagram of an all-fibre phase modulator.

The light phase propagating through an optical fibre is defined as $\phi = \beta\, L = k_0\, n_e\, L$, and when the fibre is subjected to modulating effects an optical phase shift $\Delta\phi$ occurs:

$$\Delta\phi = \beta \Delta L + L\Delta\beta = k_0\, n_e\, L\, s_z + k_0\, L\, \Delta n_e \qquad (3.26)$$

where s_z is the axial strain and represents the change in length; the second term is the change in refractive index due to the photoelastic effect. If we consider now the induced uniform pressure field with a radial strain s_r, we have:

$$\Delta\varphi = k_0\, n_e\, L\{s_z - (n_e^2/2)[(p_{11}+p_{12})s_z + p_{12}s_z]\} \qquad (3.27)$$

where p_{ij} are the photoelastic coefficients. This phase modulation has been demonstrated over a frequency range of 20–50 MHz with a phase sensitivity of 3.5×10^{-6} rad/(V/m). A plastic sheath 50–100 μm thick is used (Imai et al. 1987). A fibre-optic magnetic force modulator using an aluminium sheathed current-carrying fibre coil has been demonstrated by Godil (1989).

Another method of constructing a simple modulator is to use a fibre loop attached to a piezoelectric plate, which produces two strains in opposite directions, leading to a standing wave and resonance frequencies (Zervas and Giles 1988). Another type of modulator is based on the optical Kerr effect, which induces a change in refractive index by means of a high-intensity light beam. This is described in Chapter 5.

3.8.2 Intensity modulators

Using a taper made on a single-mode fibre length and exciting an acoustic wave in the taper waist with a PZT disc, the input single mode is coupled to the second mode, which is non-guided and is stripped by the fibre coating at the end of the device. The optical output amplitude varies with acoustic amplitude for a given wavelength. The resonant wavelength can be tuned by changing the acoustic frequency. This device can thus also be used as a tunable filter (Birks et al. 1994).

3.9 LOOPS AND RINGS

By using fibre couplers, several devices, such as: rings, loops and mirrors (Figure 3.39) can be made. They can be combined in order to construct specific components for a number of applications (Ja 1991).

3.9.1 Resonators

A fibre reflector can be made by forming a fibre loop between the output ports of a low loss fibre coupler (see Figure 3.39c) (Capmany and Muriel 1990, Mortimore 1988, Levy 1992Ya 1990). It is made by fused taper technology from a single length of fibre, thus reducing unwanted reflections and additional losses caused by fibre splices. The reflectance of the fibre loop depends upon the coupling characteristics of the directional coupler and the degree of birefringence in the fibre loop. The wavelength response is dependent upon the degree of birefringence in the loop and the spectral characteristics of the coupler. The light may for example, be split equally by the coupler to form two fields. 50% of the input light travels clockwise round the loop and 50% anticlockwise. Light coupled from port 1 to port 4 across the waveguide suffers a $\pi/2$ phase lag. The transmitted intensity in port 2 is the sum of the clockwise and the anticlockwise field of phase $-\pi$, both of equal amplitude. This results in zero transmitted intensity at port 2, and conservation of energy requires that all input light is reflected back along the input port 1. This description is simplified: it ignores the effects of birefringence of the loop and polarization response, which are examined in Chapter 4.

These reflectors are used as passive Fabry-Pérot devices, or to form a resonant cavity for a rare-earth doped laser fibre (see Chapter 6) (Miller *et al.* 1987, Millar *et al.* 1988).

3.9.2 Delay lines and circulators

Loop reflectors (see Figure 3.39c) are also used as delay line memories (Thompson and Giordano 1987, Thompson 1988) in time slot interchangers. These architectures permit interchange data, i.e., data are stored in a loop as they are received and can be recovered by random access. These loops have the same physical fibre length but are read at different delay times corresponding to different circulation

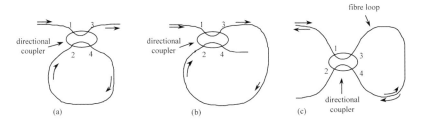

Figure 3.39. Schematic of a: (a) fibre ring, (b) fibre loop, (c) loop reflector.

numbers in delay lines. These architectures offer applications in computing and signal processing. A ring with two couplers and a length of active erbium doped fibre has been tested for use as a delay line fibre-optic filter in high-speed signal processing (Pastor *et al.* 1995).

A third application of delay line loops is the measurement of laser frequency fluctuation. The analysis is achieved by a delayed self-heterodyning set-up using an acousto-optic modulator for the local signal, the delay fibre and an erbium doped fibre amplifier being in the same loop (Ishida 1992).

3.10 INTERFEROMETERS

The principle of an interferometer is to divide a monochromatic beam into two paths and recombine them on a detector. Each path has a specific phase and the phase difference is transformed by the detector into intensity modulation.

The phase of a light wave propagated in an optical fibre is more sensitive to external influences than any other propagation parameters. Optical fibre sensors utilize this phase dependence and give a high sensitivity. There are two general approaches in the development of fibre optic interferometric sensors (Culshaw 1984, Jeunhomme 1983, Dakin and Culshaw 1988, Dakin and Culshaw 1989, Jones 1987). The first relies on the disturbance phase and is normally called an interferometric sensors (see Figure 3.40). The fibre-optic Mach-Zehnder interferometer utilizes two single-mode fibres in the arms of the interferometer to detect the relative phase shift of light (see Figure 3.40a). An all single-mode fibre interferometer is made with pigtailed laser diodes and fibre optics couplers (Schmuck and Strobel 1986) (see Figure 3.40b). The second approach considers the relative phase displacement of two polarization eigenmodes (such devices are called polarimetric sensors) (Rogers 1985, Mermelstein 1986) (see Figure 3.40c), or of the two first modes HE_{11} and HE_{21} (Lacroix *et al.* 1988).

The interferometer output intensity is given by the superposition of the two transmitted fields. If we have in each arm respectively:

$$\tilde{E}_1 = E_{01} \, e^{i(\omega t - \phi_1)}$$
$$\tilde{E}_2 = E_{02} \, e^{i(\omega t - \phi_2)}$$
(3.28)

Figure 3.40. Interferometer: (a) Mach-Zehnder interferometer, (b) optical fibre Mach-Zehnder interferometer, (c) the same with one arm, (d) heterodyne method with a Michelson interferometer, (e) Sagnac interferometer.

Interferometers 119

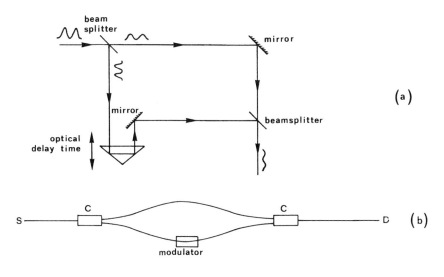

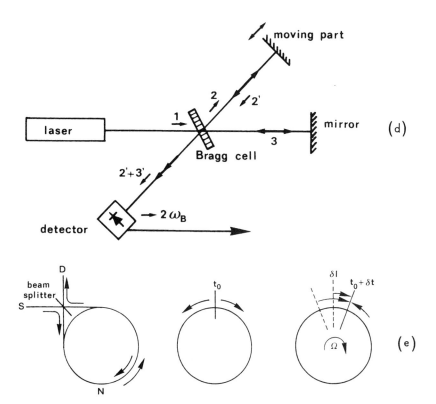

where ϕ_1, ϕ_2 are the phases in each arm given by $\varphi_i = (2\pi n_e l_i)/\lambda_0$, l_i is the fibre path length on the arm i, and n_e the effective index ($n_e \approx n_1$ index of the core), the resulting field is $\tilde{E} = \tilde{E}_1 + \tilde{E}_2$, and the intensity is:

$$I = \tilde{E}\,\tilde{E}^* = I_1 + I_2 + \chi(I_1 I_2)^{\frac{1}{2}}\cos(\varphi_1 - \varphi_2) \qquad (3.29)$$

* is the complex conjugate quantity and χ designates the degree of coherence between the two fields, $I_1 = \tilde{E}_1\tilde{E}_1^*, I_2 = \tilde{E}_2\tilde{E}_2^*$. In the particular case where $E_{01} = E_{02} = E_0 = (I_0)^{1/2}$, the intensity is:

$$I = 2I_0[1 + \cos\phi] \qquad (3.30)$$

$\phi = \phi_1 - \phi_2$ is the resulting phase. The variation δI in function of $\delta\phi$ is given by

$$\delta I = 2I_0 \sin\phi\,\delta\phi$$

we see that $\delta I \approx 0$ when $\sin\phi \approx 0$ i.e. if $\phi \approx 0$ (or π). In order to obtain a better result ($\phi = \pi/2$) a phase modulator is inserted in the reference arm. This is called *homodyne detection* (see Figure 3.40b).

A second method of measurement is the heterodyne method. Many of the problems with interferometric detection can be overcome by the use of a heterodyne interferometer. In such a device the reference beam is frequency shifted by using a Bragg cell, imposing a frequency shift ω_B on the deflected beams (see Figure 3.40d). The transmitted waves through the different arms are

1     $E_0 e^{i\omega t}$

2     $E_0 e^{i[(\omega+\omega_B)t]}$

2' $E_0 e^{i[(\omega+\omega_B)t-\phi(t)]}$

3     $E_0 e^{i(\omega t-\psi)}$

3' $E_0 e^{i[(\omega-\omega_B)t-\psi]}$

ψ is the phase difference due to the way 3. If the amplitudes in the two arms are equal, the output from the detector is:

$$I = 2E_0^2[1 + \cos(2\omega_B t - \phi(t) + \psi)]$$

The frequency is well outside the 1/f noise region. So a gain in achievable signal-to-noise ratio of over 20 dB is obtained.

In a Sagnac interferometer, the two beams issued from the beamsplitter are injected into a single fibre-optics coil (see Figure 40e). The beams circulate in the coil with clockwise and anticlockwise propagations respectively. The phase difference $\delta\phi$ is due to a small rotation of the coil around it's axis. $\delta\phi$ is related to the delay time δt measured in the fixed system of reference. This principle finds practical applications in accelerometer systems and non-mechanical gyroscopes.

Polarization-maintaining single-mode (PMSM) fibres with elliptical cladding have been developed in order to maintain linear polarization over long distances. An additional advantage lies in their two orthogonal independent optical paths, which are along their principal axes, a fast axis and a slow axis. The two paths make up two arms of an interferometer, and the beat length between two normal modes changes with strain (see Chapter 4 § 4.5).

A two-port all-fibre reflection Mach-Zehnder interferometer consists of two directional couplers configured so that there are only two open ports. The light beam undergoes splitting and coherent recombinaison into the two couplers (see Figure 3.41) (Millar et al. 1989). The light exits the interferometer either as a transmitted output from the opposite port or as a reflected output from the launching port. The entire device can in principle be made from one continuous length of fibre. When the two couplers have coupling ratio of $CR_i = 0.5$, the ratio of the intensities at the output I_0 and at the input I_1 is given by:

$$(I_0/I_1)_i = [(B_i + D_i)^2 - 4B_iD_i \sin^2(\delta\phi)] \exp(-2\alpha L_3)$$

$\delta\phi = \beta (L_1 - L_2)$ is the phase difference between the two arms of lengths L_1 and L_2. B_i, D_i, (in which i = 1, 2 depending on whether the output is

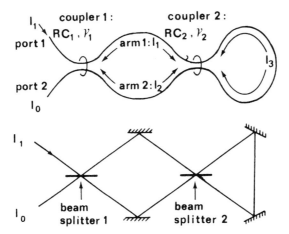

Figure 3.41. (a) Fibre reflection Mach-Zehnder interferometer, (b) equivalent interferometer bulk optical mirrors and beam-splitters [After Millar et al. 1989].

from port 1 or 2) are expressed as functions of the coupling losses γ_i, of the coupling ratio CR_i and of the transmitance t_i and reflectance r_i of the couplers; l_3 is the length of the loop, and α is the absorption coefficient of the fibre (Millar et al. 1989).

When the losses of the fibre and of both the couplers are negligible, the intensity equations can be written in the simple form:

$$(I_0/I_1)_i = \sin^2(\delta\phi) \qquad i = 1 \qquad \text{if} \qquad r_2 = 1$$
$$= \cos^2(\delta\phi) \qquad i = 2 \qquad \text{if} \qquad r_1 = 1 = r_2$$

This interferometer can be potentially useful as a sensor or a reflection modulator (Capmany and Muriel 1990).

REFERENCES FOR CHAPTER 3

Agarwal AK 1985 *Fibre and Integrated Optics* **6** 27–53
Agrawal GP and Radic S 1994 *IEEE Photonics Technol. Lett.* **6** 995–7
Ankiewicz A, Snyder AW and Zheng XH 1986 *J. Lightwave Technol.* **4** 1317–23
Annovazzi-Lodi V and Donati S 1990 *J. Opt. Comm.* **11** 107–21
Arkwright J, Chu PL and Tjugiarto 1993 *IEEE Photonics Technol. Letters* **5** 1216–18
Ball GA, Morey WW and Waters JP 1990 *Electron. Lett.* **26** 1829–30
Ball GA, Glenn WH and Morey WW 1994 *IEEE Photonics Technol. Lett.* **6** 741–3
Belovolov MI, Dianov EM, Kuznetzov AV and Prokhorov AM 1987 *Fibre and Integrated Optics* **6** 239–53
Bennion I, Reid DCJ, Rowe CJ and Stewart WJ 1986 *Electron. Lett.* **22** 341–3
Bergh RA, Kotler G and Shaw HJ 1980 *Electron. Lett.* **16** 260–1
Bilodeau F, Johnson DC, Thiérault S, Malo B, Albert J and Hill KO 1995 *IEEE Photonics Technol. Lett.* **7** 388–90.
Birks TA and Hussey CD 1988 *Opt. Lett.* **13** 681–3
Birks TA, Russel PStJ and Pannell CN 1994 *IEEE Photonics Technol. Lett.* **6** 725–7
Birks TA, Russel PStJ and Culverhouse D O 1996 *J. Lightwave Technol.* **14** 2519–29
Blake JN, Kim BY and Shaw HJ 1986 *Opt. Lett.* **11** 177–9
Blake JN, Kim BY, Engan HE and Shaw HJ 1987 *Opt. Lett.* **12** 281–3
Boucouvalas AC and Georgiou G 1987 *IEE Proc. Pt.J.* **134** 191–5
Brackett CA *1990 IEEE J. on selected area in comm.* **8** 948–64
Brady GP, Hope S, Lobo Ribeiro AB, Webb DJ, Reekie L, Archambault JL and Jackson DA 1994 *Optics Comm.* **111** 51–4

Bures J, Lacroix S and Lapierre J 1983 *Appl. Opt.* **22** 1918–23
Bures J, Lacroix S, Veilleux C and Lapierre J. 1984 *Appl. Opt.* **23** 968–9
Capmani J and Muriel MA 1990 *J. Lightwave Technol.* **8** 1904–19
Cassidy SA and Yennadhiou P 1988 *IEEE J. on Selected Areas in Communications* **6** 1044–50
Chattopadhyay T and Nakajima M 1990 *Optics Comm.* **77** 144–6
ChenYH, Chang CH, Yang YL, Kuo IY and Liang TC 1999 *Optics Comm.* **169** 245–62
Chiang KS 1987 *Opt. lett.* **12** 431–3
Culshaw B 1984 *Optical fibre sensing and signal processing*(Peter Peregrinus Ltd)
Dakin JP and Culshaw B 1988 *Optical fibres sensors I Principles and components* (Artech House: London)
Dakin JP and Culshaw B 1989 *Optical fibres sensors II Systems and applications* (Artech House: London)
De Fornel F, Ragdale CM and Mears RJ 1984 *IEE Proc. PtH* **131** 221–8
De Paula RP and Moore EL 1984 *SPIE Proceedings* **478** 3–11
Digonnet MJ F and Shaw HJ 1982 *IEEE J. Quantum Electron.* **18** 746–53
Digonnet MJF, Feth JR, Stokes LF and Shaw HJ 1985 *Opt. Lett.* **10** 463–5
Dong L, Berkey GE, Chen P and Weidman DL 2000 *J.Lightwave Technol.* **18** 1018–23
Dragone C 1989 *J. Lightwave Technol.* **7** 479–89
Dragone C, Henry CH, Kaminow IP and Kistler RC 1989 *IEEE Photon. Technol. Lett.* **1** 241–3
Eisenmann M and Weidel E 1988 *IEEE J. Lightwave Technol.* **6** 113–19
Eriksson U, Blixt P and Tellefsen JA 1994 *Optics Lett.* **19** 1028–30
Falciai R, Scheggi AM and Schena A 1990 *Intern. J. optoelectronics* **5** 41–6
Fertein E, Legoubin S, Douay M, Canon S, Bernage P, Niay P, Bayon F and Georges T 1991 *Electron. Lett.* **27** 1838–9
Geckeler S 1990 *IEEE J. on selected areas in comm.* **8** 1115–19
Georgiou G and Boucouvalas AC 1985 *IEE Proc. Pt J.* **132** 297–302
Godil AA 1989 *J. Lightwave Technol.* **7** 2052–4
Goldburt ES and Russel PStJ 1986 *Opt. Lett.* **11** 51–2
Griffin R, Love JD, Lyons PRA, Thorncraft DA and Rashleigh SC 1991 *IEEE J. of Lightwave Technol.* **9** 1508–17
Gusarov A, Hong Ky N, Limberger H G, Salathé R P and Fox G R 1996 *J. Ligthwave Technol.* **14** 2771–7
Hwang J K, Yun S H and Kim B Y 1999 *Optics Lett.* **24** 1263–65
Heffner BL and Kino GS 1987 *Opt. Lett.* **12** 208–10

Heismann F and Ulrich R 1984 *Appl. Phys. Lett.* **45** 490
Hill AM 1986 *J. Lightwave Technol.* **4** 785–9
Hill KO, Malo B, Johnson DC and Bilodeau F 1990 *IEEE Photonics Technol. Lett.* **2** 484–6
Hogari K and Matsumoto T 1990 *IEEE J. Lightwave Technol.* **8** 722–7
Huang HS and Chang HC 1990a *IEEE J. Lightwave Technol.* **8** 823–31
Huang HS and Chang HC 1990b *IEEE J. Lightwave Technol.* **8** 832–7
Hwang I K, Yun S H and Kim B Y 199 *Optics Lett.* **24** 1263–5
Imai M, Shimizu T, Ohtsuka Y and Odajima A 1987 *IEEE J. Lightwave Technol.* **LT5** 926–31
Imai M, Fujiwara S, Ohtsuka Y and Odajima A 1988 *Opt. Lett.* **13** 838–40
Iocco A, Lumberger HG, Salathè RF, Everall LA, Chisholm KE, Williams JAR and Bennion I 1999 *J. Lightwave Technol.* **17** 1217–21
Irshid MI and Kavehrad M 1991 *IEEE Photonics Technol. Lett.* **3** 247–9
Irshid MI and Kavehrad M 1992 *IEEE Photonics Technol. Lett.* **4** 58–60
Ishida O 1992 *IEEE Photonics Technol. Lett.* **4** 1304–7
Ja YH 1990 *Optics Commun.* **75** 239–45
Ja YH 1991 *J. of Optical Commun.* **12** 29–32
Jackel JL, Hackwood S, Veselka JJ and Beni G 1983 *Appl. Opt.* **22** 1765–70
Jacob-Poulin AC, Vallée R, La Rochelle S, Faucher D and Atkinson GR 2000 *J. Lightwave Technol.* **18** 715–20
Jauncey IM, Reekie L, Mears RJ, Payne DN, Rowe CJ, Reid DcJ, Bennion I and Edge C 1986 *Electron. Lett.* **22** 987
Jen CK and Goto N 1989 *J. Lightwave Technol.* **7** 2018–23
Jeunhomme LB 1983 *Single-mode fibre optics: Principles and applications* (Marcel Dekker: New York)
Johnson DC, Hill KO, Bilodeau F and Faucher S 1987 *Electon. Lett.* **23** 668
Jones BE (ed) 1987 *Current advances in sensors* (Adam Hilger: Bristol)
Karlsson M 1994 *Optics Communications* **112** 48–54
Kashyap RK 1999 *Fibre Bragg Gratings* (Academic Press: New York)
Kawasaki BS, Hill KO and Lamont RG 1981 *Opt. lett.* **6** 327–8
Kieli M and Herczfeld PR 1986 *IEEE J. Lightwave Technol.* **4** 1729–31
Kim BY, Blake JN, Engan HE and Shaw HJ 1986 *Opt. Lett.* **11** 389–91
Kumar A, Das UK, Varshney RK and Goyal IC 1990 *J. Lightwave Technol.* **8** 34–8
Kuwahara H, Hamasaki J and Sarto S 1975 *IEEE Trans. MTT* **16** 179
Lacroix S, Gonthier F and Bures J 1986 *Opt. Lett.* **11** 671–3
Lacroix S, Gonthier F, Black RJ and Bures J 1988 *Opt. Lett.* **13** 395–7
Lamouroux B and Prade B 1987 *J. Opt. Soc. Am.* **4** 327
Lamouroux B, Morel P, Prade B and Vinet J 1985 *J. Opt. Soc Am* **2** 759

Legoubin S, Fertein E, Douay M, Bernage P, Niay P, Bayon F and Georges T 1991 *Electron. Lett.* **27** 1945–6
Leminger O and Zengerle R 1987 *Opt. lett.* **12** 211–13
Leminger O and Zengerle R 1990 *IEEE J. Lightwave Technol.* **8** 1289–91
Levy JF 1992 *IEE Proc. PtJ* **139** 313–17
Liao FJ and Boyd JT 1981 *Appl. Opt.* **20** 2731–4
Liaw SK, Ho KP, Lin C and Chi S 1999 *Opt. Comm.* **169** 75–80
Liaw JW, Chen YK and Guo WY 1993 *Intern. J. Optoelectronics* **8** 21–32
Loeb ML and Stilwell GR 1990 *J. Lightwave Technol* **8** 239–42
Liu W F, Russel PStJ and Dong L 1998 *J. Ligthwave Technol.* **16** 2006–9
Liu W F, Liu I M, Chung L W, Huang D W and Yang C C 2000 *Optics Lett.* **25** 1319–21
Lu KW, Eiger MI and Lemberg HL 1990 *IEEE J on selected areas in comm.* **8** 1058–67
Mahlein HF 1983 *Fibre and Integrated Optics* **4** 339–72
Marcuse D 1985 *Electron Lett* **21** 726
Martini G 1987 *Opt. and Quant. Electron.* **19** 179–90
McLandrich MN, Orazi RJ and Marlin HR 1991 *J. Lightwave Technol.* **9** 442–7
Meltz G, Morey WW and Glenn WH 1989 *Opt. Lett.* **14** 823–5
Mermelstein MD 1986 *IEEE J. Lightwave Technol.* **LT4** 449–53
Miles E 1992 *Fibre and Integrated Optics* **10** 323–50
Millar CA, Brierley MC and Mallinson SR 1987 *Opt. Lett.* **12** 284–6
Millar CA, Miller ID, Mortimore DB, Ainslie BJ and Urquhart P 1988 *IEE Proc Pt J.* **135** 303.
Millar CA, Harvey D and Urquhart P 1989 *Opt. Comm.* **70** 304–8
Miller ID, Mortimore DB, Urquhart P, Ainslie BJ, Craig SP, Millar CA and Payne DB 1987 *Appl. Opt.* **26** 2197–201
Mizunami T, Djambova TV, Niiho T and Gupta S 2000 *J. Lightwave Technol.* **18** 230–5
Morey WW, Dunphy JR and Meltz G 1992 *Fibre and Integrated Optics* **10** 351–60
Morishita K 1991 *J. Lightwave Technol.* **9** 584–9
Mortimore DB 1988 *IEEE J. Lightwave Technol.* **6** 1217–24
Niay P, Bernage P, Legoubin S, Douay M, Xie WX, Bayon JF, Georges T, Monerie M and Poumellec B 1994 *Optics Commun.* **113** 176–92.
Niay P, Bernage P, Taunay T, Douay M, Delevaque E, Boj S and Poumellec B 1995 *IEEE Photonics Technol. Lett.* **7** 391–3.
Noda J, Okamoto K and Yokohama I 1987 *Fibre and Integrated optics* **6** 309–30

Ohshima S, Ito T, Donuma KI, Sugiyama H and Fujii Y 1985 *IEEE J. Lightwave Technol.*LT **3** 556–60
Okamoto K 1990 *J. Lightwave Technol.* **8** 678–83
Ortega B and Capmany J 1999 *J. Lightwave Technol.* **17** 1241–7
Ortega B and Dong L 1999 *J. Lightwave Technol.* **17** 123–8
Ozeki T and Kawasaki BS, 1976 *Electron. Lett.* **12** 151–2
Parriaux O, Bernoux F and Chartier G 1981a *J. Opt. Commun.* **2** 105–9
Parriaux O, Gidon S and Kuznetsov AA 1981b *Appl. Opt.* **20** 2420–3
Pastor D, Sales S, Capmany J, Marti J and Cascon J 1995 *IEEE Photonics Technol. Lett.* **7** 75–7
Patterson DB, Howell MD, Digonnet M, Kino GS and Khuri Yakub BT 1990 *IEEE J. Lightwave Technol.* **8** 1304–12
Peng GD and Ankiewicz A 1991 *IEE Proc. PtJ* **138** 33–8
Poole CD, Townsend CD and Nelson KT 1991 *J. Lightwave Technol.* **9** 598–
Qian JR 1986 *Electron. Lett.* **22** 304–6
Ragdale CM and Goodman SE 1985 *Proc. of the SPIE* **574** 110–14
Ragdale CM, Reid D, Robbins DJ, Buus J and Bennion I 1990 *IEEE J on selected area in commun.* **8** 1146–50
Rawson EG and Nafarrate AB 1978 *Electron. Lett.* **14** 274–5
Reid DCO, Ragdale CM, Bennion I, Robbins DJ, Buus J and Stewart WJ 1990 *Electron. Lett.* **26** 10–12
Riza N A and Sumriddetchkajorn S 1999 Optics Comm. **169** 233–44
Risk WP, Youngquist RC, Kino GS and Shaw HJ 1984 *Opt. Lett.* **9** 309–11
Risk WP, Kino GS, Khuri-Yakub 1986 *Opt. Lett.* **11** 578–80
Rocks M 1987 *J. Opt. Comm.* **8** 22–4
Roeksabutr A and Chu P L 1996 *J. Lightwave Technol.* 14 2362–6
Rogers AJ 1985 *IEE Proc. Part J.* **132** 303
Rowe CJ, Bennion I and Reid DCJ 1987 *IEE Proc. Pt J.* **134** 197–202
Russel P St J and Ulrich R 1985 *Opt. Lett.* **10** 291–3
Saleh AAM and Kogelnik H 1988 *IEEE J. Lightwave Technol.* **6** 392–8
Savin S, Digonnet MJF, Kino GS and Shaw H J 2000 *Optics Lett.* **25** 710–12
Schmuck H and Strobel O 1986 *J. Opt. Comm.* **7** 86–91
Shankar NK, Morris JA, Yakymyshyn CP and Pollock CR 1990 *IEEE Photonics Technol. Lett.* **2** 147–9
Sharma A, Kompella J and Mishra PK 1990 *J. Lightwave Technol.* **8** 143–51
Sheem SK and Galliorenzi TG 1979 *Opt. Lett.*4 29–31
Shi CX 1992 *IEEE Photonics Technol. Lett.* **4** 1279–81
Shin JD, Lee CE, Conway DB, Atkins RA and Taylor HF 1989 *IEEE Photon. Technol. Lett.* **1** 276–7

Shipley SP, Georgiou G, Boucouvalas AC 1987 *IEE Proc. PtJ* **134** 203–7

Snyder A and Love JD 1983 *Optical Waveguide Theory* (Chapman and Hall: New York)

Sorin WV, Kim BY and Shaw HJ 1986 *Opt. Lett.* **11** 581–3

Suzuki Y and Kashiwagi H 1976 *Appl. Opt.* **15** 2032–3

Tabiani M and Kavehrad M 1991 *J. Lightwave Technol.* **9** 448–55

Takeuchi Y and Noda J 1992 *IEEE Photonics Technology Lett.* **4** 465–7

Thompson RA 1988 *IEEE J. Select Areas Commun* **6** 1096–106

Thompson RA and Giordano PP 1987 *IEEE J. Lightwave Technol.* LT **5** 154–62

Tomlinson WJ 1980 *Appl. Opt.* **19** 1127–38

Town GE, Sugden K, Williams JAR, Bennion I and Poole SB 1995 *IEEE Photonics Technol. Lett.* **7** 78–80

Van Roey J, Van Der Donk J and Lagasse PE 1981 *J. Opt. Soc. Am.* **71** 803–10

Van Wiggeren GD, Gaylord T K, Davis D D, Braiwish M, Glytsis E N and Anemogiannis E 2001 *Optics Lett.* **26** 61–3

Wang CC 1987 *J. Lightwave Technol.* **5** 77–81

Whalen MS and Walker KL 1985 *Electron. Lett.* **21** 724

Williams JAR, Bennion I and Doran NJ 1995 *Optics Communications* **116** 62–6

Willner AE, Saleh AMM, Presby HM, Di Giovanni DJ and Edwards CA 1991 *IEEE Photonics Technol. Lett.* **3** 250–2

Wilson SJ and Bricheno T 1990 *Opt. Comm.* **75** 106–10

Winzer G, Mahlein HF and Reichelt A 1981 *Appl. Opt* **20** 3128–35

Wright JV 1986 *Electron. Lett.* **22** 320–1

Wright JV, Mallinson SR and Millar CA 1988 *IEEE J. selected area comm.* **6** 1160–8

Wysocki P, Kostenbauder AG, Kim BY and Siegman AE 1989 IEEE *J. Lightwave Technol* **7** 1964–6

Yanagawa H, Nakamura S, Ohyama I and Ueki K 1990 *J. Lightwave Technol.* **9** 1292–7

Yijiang C 1989 *Opt. and Quantum electron.* **21** 491–8

Youngquist RC, Brooks JL, Risk WP, Kino GS and Shaw HJ 1985 *IEE Proc. Part J* **132** 277–86

Zengerle R and Leminger OG 1985 *J. Opt. Commun.* **4** 150–2

Zervas MN and Giles IP 1988 *Opt. Lett.* **13** 404–6

Zheng XH and Snyder AW 1987 *Electron. Lett.* **23** 182–4

EXERCISES

3.1 – In a 2×2 coupler made with two identical fibres the beat length has a value $L_b = 500\,\mu m$. A power $P_1 = 100$ mW is injected into port 1. Calculate the coupling coefficient C. What is the power in port 2 for $z = 250\,\mu m$, $z = 2.5$ mm.

3.2 – For a phase shifter:
    a- Establish the value of $\delta\varphi$ versus temperature variation δT
    b- Calculate $\delta\varphi$ if $n_1 = 1.46$, $L = 1$m, $\lambda = 1\mu m$, $\delta T = 1°C$, $dn_1/dT = 0.68 \times 10^{-5}$ /°c, $\alpha = 5.5 \times 10^{-7}$ /°c.

3.3 – A Bragg grating with a grating period Λ is obtained by interferometric exposure into the fibre core. Demonstrate that is acts as a mirror for a wavelength λ_B.
What is the value of λ_B if $\Lambda = 531$ nm and $n_1 = 1.46$.
Show that this device can be used as a temperature sensor.

3.4 – A fibre Bragg grating has a period $\Lambda = 4\,\mu m$. What are the diffracted orders when the fibre propagates light at wavelength $\lambda = 0.8\,\mu m$ ($n_e = 1.46$).

3.5 – A gyroscope is an optical fibre Sagnac interferometer. A single mode fibre is wound into N loops of radius R (see figure 3.40).
    a – What is the phase shift $\delta\varphi$ if the device rotates at an angular velocity
    Ω (rd / s).
    b – Calculate $\delta\varphi$ if $\lambda = 1.5\,\mu m$, $N = 800$, $R = 100$ mm, $c = 3 \times 10^8$ m/s,
    $\Omega = 10^{-4}$ rd / s, $n = 1.4$.

Chapter 4

DEVICES USING POLARIZED LIGHT

Multimode and single-mode fibres have been developed to a high degree of performance for communication and sensor applications. Single-mode optical fibre has applications not only in very wide bandwidth optical communication systems, but also in current rotation monitoring and other interferometric devices. For such applications the fibres operate with coherent polarized light. In order to obtain measurements of the desired accuracy it is often essential that the light has only one polarization or that the state of polarization is controlled.

Birefringent fibres maintain two linear orthogonal polarizations along their length (see Chapter 1 § 1.4). Polarization maintaining and polarization splitting couplers, polarizers, depolarizers, wavelength devices, isolators, circulators and interferometers, are all currently manufactured as fibre optics components.

4.1 POLARIZATION IN SINGLE-MODE FIBRES

In many applications, the state of polarization of the modes in a fibre needs to be strictly controlled, for example in interferometric sensors, in coherent transmittance, or in coupling to integrated optical circuits. In ordinary single-mode fibres, the state of polarization is undeterminated. As seen in Chapter 1, in an unperturbed single-mode fibre, two degenerate eigenpolarization modes HE_{11}^{x} and HE_{11}^{y}, whose electric fields can be denoted as e_x and e_y, can be propagated. Theoretically, when a perfect (i.e., circularly symmetric) fibre is laid in a straight line, linearly-polarized light launched at the input will maintain this state along the fibre length to the output. In practice, fibres cannot be made as perfectly cylindrical structures. There are intrinsic imperfections, bends, stresses, vibrations and changes in temperature, which effects produce inhomogeneities. When uniform perturbations are present in a fibre, elliptical birefringence is introduced. This leads to different propagation constants of the eigenpolarization modes and degeneracy is no longer present. The resultant electrical field E is then represented as a linear superposition of the fields e_x and e_y of the unperturbed fibre.

$$E = A_x(z)e_x + A_y(z)e_y \tag{4.1}$$

where $A_x(z)$ and $A_y(z)$ are the amplitude of the eigenmodes, depending on the position z along the fibre. The change of the amplitudes along the fibre is described by the coupled mode equations:

$$\frac{d}{dz}\begin{bmatrix} A_x(z) \\ A_y(z) \end{bmatrix} = i \begin{bmatrix} A_{11} & A_{12} \\ A_{21} & A_{22} \end{bmatrix} \begin{bmatrix} A_x(z) \\ A_y(z) \end{bmatrix} \tag{4.2}$$

where A_{jk} are the coupling coefficients and $A_{21} = A_{12}{}^*$. When a linearly polarized light beam is launched, it may be decomposed by the fibre into two linearly polarized orthogonal components along the two principal transverse axes with different phase velocities. Thus coupling between these two components will cause the state of polarization to vary randomly along the length of the fibre. In single-mode fibre, this modal birefringence due to geometrical core deformation and external stress gives rise to polarization mode dispersion (Hakli 1996). It may cause significant problems in multigigabit systems using direct detection techniques as well as in coherent detection systems. A review of polarization mode dispersion measurement methods in long optical fibres is given by Namihira and Wakabayashi (1991).

Introducing strong linear birefringence into the fibre reduces the amount of coupling between the two mode components and stabilizes the linear polarization state. As seen in Chapter 1, several methods are available. One method is to make the core non-circular in shape, so that the refractive index distribution in the two principal directions differs, i.e., a fibre with an elliptical core. Another method of producing linear birefringence is to introduce an asymmetric stress over the core of the fibre. The core and cladding remain circular, but asymmetric sectors of a different expansion coefficient are introduced into the cladding: 'bow-tie' or 'Panda' fibres (see Chapter 1 Figure 1.11). These fibres are termed *polarization-maintaining fibres*. In case of 'bow-tie' fibres, beat lengths of less than 1 mm (modal birefringence $B = 6 \times 10^{-4}$) can be obtained.

A fibre exhibiting a high degree of linear birefringence as described in Chapter 1 and, as indicated above, can operate in distinct ways. Assuming that the two orthogonal modes have low transmittance losses and propagate with the same attenuation, then when an equal amount of light is launched into each of the modes, the phase constants are different and coupling results. So the state of polarization changes periodically along the length of the fibre, from linear to circular and back again. However, when only one of the modes is launched and no mode conversion occurs, the light continues to be linearly polarized along the length of the fibre. Under external perturbation (bending, stress, etc.) some of the first polarization will couple into the orthogonal mode and will propagate in that mode through to the output.

The development of linearly birefringent optical fibres has led to the construction of highly anisotropic optical fibres in which differential polarization mode attenuation may be present under certain conditions. In the case of these kinds of fibre one method of operating is to introduce attenuation preferentially into one of the modes. When the light is launched in the low loss mode, it will continue in that mode along the length of the fibre. When the light is launched in the orthogonal mode, with high loss, it will be attenuated. In practice this means that single polarization is achieved when one polarization state of the fundamental mode, for example the y-polarization, is guided and the other is leaky. So the output remains linearly polarized despite the mode coupling for any state of input polarization. Such a fibre is termed a single polarization fibre. For example, in fibres exhibiting a polarizing effect the propagation difference $\delta\beta$ is 6×10^3 rad/m (beat length $L_b = 1$ mm) the losses are $\alpha_x = 5$ dB/km and $\alpha_y = 40$ dB/km i.e. $\Delta\alpha = 35$ dB/km. The practical importance of these fibres is related to the construction of fibre optic sensors and line polarizers.

Theoretical analysis of single polarization fibres has been mainly concentrated upon mechanisms which achieve polarizing properties. The polarization properties depend on bending loss difference between the two polarization modes, the macro and microbending effects in 'bow-tie' anisotropic fibres and the stress-induced birefringence in PANDA fibres. In all cases the birefringence and the differential attenuation are of prime interest. In order to introduce a preferential loss into one mode of a bow-tie fibre, the fibre is wound into a coil. So there are different refractive index distributions in the two principal transverse planes and the bending loss edges of the two modes of the bow-tie fibre will be on different wavelengths. The attenuation of the two modes is very different in a wavelength region. The spectral dependence of the attenuation coefficients for the bow-tie single-mode fibre is given in Figure 4.1.

For enhancing birefringence of elliptical single-mode fibres the use of an azimuthally modulated index in the inner layer of an elliptical three-layer fibre is attractive (Cancielleri *et al.* 1990). Theory of these kinds of fibres can be found in several books and papers, e.g., Kortenski *et al.* (1990) Ulrich and Simon (1979). The evolution of the state and

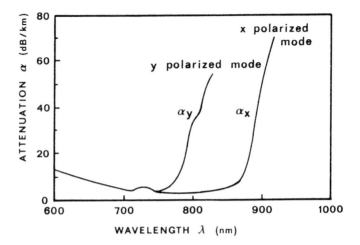

Figure 4.1. Spectral dependence of the attenuation coefficients and for 'bow-tie' single-mode fibre [After Varnham et al. 1983].

degree of polarization in uniformly anisotropic single-mode and single-polarization optical fibre in the quasi-monochromatic case is obtained using the Muller-Stokes matrix formalism. Starting from equation 4.1 the Stokes parameters can be expressed in terms of the amplitudes $A_x(z)$ and $A_y(z)$ as follows:

$$\begin{aligned} S_0(z) &= |A_x(z)|^2 + |A_y(z)|^2 \\ S_1(z) &= |A_x(z)|^2 - |A_y(z)|^2 \\ S_2(z) &= 2Re\left[A_x(z)A_y^*(z)\right] \\ S_3(z) &= -2\,Im\left[A_x(z)A_y^*(z)\right] \end{aligned} \qquad (4.3)$$

The Muller matrix $M(z)$ of a polarizing single-mode fibre which transforms the input Stokes vector $S^\circ = S(0)$ into the output vector is

$$S(z) = M(z)\,S(0) \qquad (4.4)$$

where
$$S(z) = \{S_0(z),\ S_1(z),\ S_2(z),\ S_3(z)\}$$

and
$$S(0) = S_0^\circ = \{\,S_0^\circ,\ S_1^\circ,\ S_2^\circ,\ S_3^\circ\}$$

For a polarizing single-mode fibre we have:

$$M^0(z) = e^{-\alpha z} \begin{bmatrix} \cosh(\Delta\alpha z) & \sinh(\Delta\alpha z) & 0 & 0 \\ \sinh(\Delta\alpha z) & \cosh(\Delta\alpha z) & 0 & 0 \\ 0 & 0 & \cos\delta\beta z & \sin\delta\beta z \\ 0 & 0 & -\sin\delta\beta z & \cos\delta\beta z \end{bmatrix} \quad (4.5)$$

$\delta\beta$ is the birefringence, $\Delta\alpha(= \alpha_x - \alpha_y)$ is the differential mode attenuation and $\alpha = \alpha_x + \alpha_y$. This matrix reduces to the matrix of a linearly birefringent optical fibre if $\Delta\alpha = 0$ (i.e. $\alpha_x = \alpha_y$) and to the matrix of an ideal polarizer if $\alpha_x = 0$ (or if $\alpha_y = 0$) when α_y (or α_x) is large.

Fibres exhibiting a high degree of circular birefringence have been made. One method of producing circular birefringence is to twist an optical single-mode fibre about its longitudinal axis. The propagation constants of modes polarized in the left and right circular directions are different.

To calculate the coupling coefficients A_{jk} of equation 4.2 resulting from a twist, it is assumed that the elastic properties and the elasto-optic tensor p_{jk} are uniform throughout the fibre. The twist τ causes a rotation $\gamma = \tau z$ of the cross section plane z. A positive value of parameter τ means a right-handed twist. For weakly guiding fibres of arbitrary index profile the influence of twist gives:

$$A_{11} = A_{22} = 0 \quad A_{12} = -A_{21} = -i\, n^2\, p_{44}\, \tau/2$$

where n is the mean refractive index of the fibre. A_{jk} coefficients cause circular birefringence or optical activity. A fibre subjected to right-handed twist τ exhibits a strain-induced rotatory optical activity of angle $\theta = g\tau$. The proportionality factor g equals $-n^2 p_{44}$; for weakly doped silica fibre, $p_{44} = (p_{11} - p_{12})/2 = -0.075$ and $n = 1.46$ for silica fibre, thus $g = 0.16$ (Ulrich and Simon 1979). Twisted fibres may be used as polarization rotators.

Another method is to produce a fibre in which the core follows a helical path around the longitudinal fibre axis.

Birefringence in the fibre can also be caused by the application of magnetic or electric fields. A magnetic field with a component H_z along the direction of propagation in the fibre produces a rotation θ of linear polarization (Faraday effect).

$$\theta = V_D \int \mathbf{H}_z\, \mathbf{z}\, dz \quad (4.6)$$

where V_D is the Verdet constant. This effect can be used in sensors for measurement of current intensity.

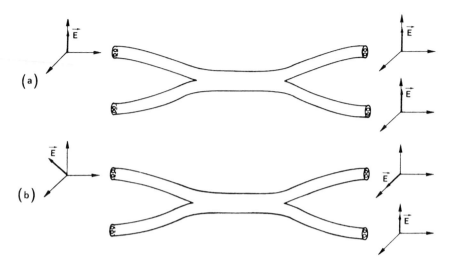

Figure 4.2. Scheme of fibre coupler realized from PANDA fibres: (a) Polarization maintaining coupler, (b) polarization splitting coupler.

4.2 X-COUPLERS USING BIREFRINGENT FIBRES

These devices are made with polarization-maintaining fibres. During the fabrication of polarization-dependent fibre couplers such as polarization-maintaining and polarization-splitting couplers, the principal axes of polarization-maintaining fibres must be parallel and precisely aligned. For example, the structure of devices fabricated with PANDA fibres is given in Figure 4.2.

Fabrication methods for fibre couplers are mainly classified (as seen for power coupler in Chapter 3) into two types: a mechanical polishing method and a fusion elongation method.

4.2.1 Polarization-maintaining couplers

Polarization-maintaining couplers have been studied by Nayar and Smith (1983), Kawachi et al. (1982), Kawachi (1983), Villarruel et al. (1983), Morishita and Takashina (1991), Yokohama et al. (1987), Ioannidis et al. (1996). The manufacture of fibre polarization-maintaining couplers uses equipment with automatic fusion–elongation processes, which is applied to fabricate single-mode couplers. With, for example, PANDA fibres, the coupling between the two fibres is caused by a decrease in the distance between the cores and in the diameters of the cores and claddings. The fabrication system is composed of two elongation stages, two pairs of microburners driven by motors and control systems. The pairs of microburners are located in the vertical and the horizontal planes and are moved longitudinally and transversely, so that the fibres are heated symmetrically. The modal birefringence of the fibre is of 1.3×10^{-4} with a refractive index difference $\Delta = 0.6\%$. The excess loss is less than 0.11 dB and a coupling ratio better than 1.3% is obtained.

Fused fibre couplers are important because of their low-cost fabrication potential and their application for optical polarization control. However, the fabrication of these devices has had a mixed success in terms of achieving a reproducing fabrication process. In particular, there are problems of fibre birefringence loss owing to fibre etching and tapering, internal axis rotation and index difference between the cladding and the stress region.

There are three origins of degradation of the cross talk in these devices: the misalignment of the principal axes (an angle of 2° gives crosstalk of -30 dB), the decrease in modal birefringence due to the mutual stress compensation produced by joining two stress-applying parts and the presence of stress-applying parts with a lower refractive index than that of the cladding (Noda et al. 1987). Dopant diffusion during the tapering of high birefringent fibres increases with initial modal birefringence (Pickett et al. 1988).

Polarization preserving couplers have been fabricated with birefringent fibres that self-align their axes within the coupling region (Pleibel et al. 1983). The birefringence of the fibre is elasto-optically introduced by a highly doped elliptical cladding. The outer shape of the fibre is made oval rather than round with dimensions $120\,\mu m \times 150\,\mu m$. When bent, the fibre bends automatically around an axis parallel to the major axis of the birefringent fibre.

All these devices show a wavelength dependence of the normalized coupling ratio between the two fibres. The two cases for the input polarized along the fast axis or the slow axis are shown Figure 4.3. The relative power in the two fibres is given by equation 3.2. In the through fibre $P_1 = \cos^2 Cz$ and in the crossed fibre $P_2 = \sin^2 Cz$, where C is the coupling coefficient and z the coupling length. C can be approximated by $C = \Omega \lambda^3$ where Ω is a constant. So the wavelength-independent part of the coupling is the product Ωz for the slow and fast modes.

4.2.2 Polarization splitting and polarization-selective couplers

In a polarization-selective directional coupler, light of only one polarization is transferred to the second fibre. The polarization

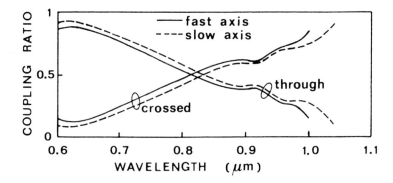

Figure 4.3. Wavelength dependence of normalized splitting ratio of polarization preserving coupler for input polarizations along both the fast and slow birefringence axes [After Pleibel et al. 1983].

selectivity arises from different mismatches for the propagation constants of the two polarizations.

The methods of fabrication originate in methods previously described (Stolen et al. 1985, Noda et al. 1987, Bricheno and Baker 1985, Yataki et al. 1985).

Birefringent fibres with elliptical cores can be held by two blocks of different bend radii. The difference in propagation constants required for polarization selectivity appears to come from differences in the depth of polishing of the two coupler halves. A holder permits transverse adjustment to maximize the coupling and longitudinal motion to study the power transfer as a function of the coupling length. The variation of power coupling as a function of coupling length is shown on Figure 4.4. Significant differences between the two coupler halves should be avoided in polarization preserving couplers. For polarization selective couplers, good polarization selectivity requires matching propagation constants for the favoured polarization with an index difference of less than 5×10^{-5}.

Polarization-splitting couplers may also be fabricated by the fusion elongation method using a micro-burner and PANDA fibres. When coupling length is long enough, polarization dependent effect is dominant, owing to the geometrical anisotropy of the fused region. Wavelength dependence of coupling ratio for PANDA fibres with a refractive index compensation of the stress-applying parts has been observed. There is a difference in the coupling ratio between X-polarization and Y-polarization. Using this difference a polarization-splitting device can be produced. The x-polarization field is coupled perfectly to the other core, whereas the y-polarization field is coupled

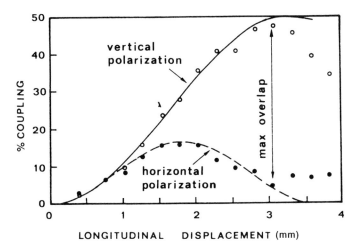

Figure 4.4. Variation in power coupling as the coupler halves are displaced longitudinally (variation of the coupling length). The solid and broken lines are calculated using a uniform coupling model with best-fit parameters [After Stolen et al. 1985].

back again to the same core (see Figure 4.5). Polarization is observed at several wavelengths.

4.3 BIREFRINGENT FIBRE POLARIZATION COUPLER

This device can couple the power between the two polarization modes of a birefringent fibre. It is based upon periodic mixing of the two polarization modes. By applying pressure to a highly birefringent fibre, an additional birefringence Δn_a can be induced:

$$\Delta n_a = S\, n^3\, KF/4b \qquad (4.7)$$

where S is a constant equal to 1.58 for a fibre of circular cross section, $2b$ is the outer diameter, n the refractive core index, $K\,(= 5 \times 10^{-12}$ SI) is the photoelastic coefficient, and F the force per unit length applied to the fibre. This birefringence is different from the intrinsic birefringence $\Delta n_i = \lambda/L_b$, where L_b is the beat length. It can be demonstrated that, when the birefringent fibre is squeezed at about 45° to the intrinsic birefringent axes, the magnitude of the total birefringence is approximately equal to the intrinsic birefringence Δn_i (Youngquist et al. 1985). The result is a rotation of the intrinsic

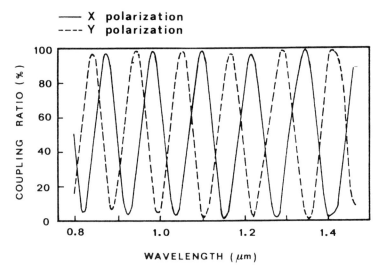

Figure 4.5. Wavelength dependence of coupling ratio for polarization-splitting coupler [After Noda *et al.* 1987].

birefringent axis without a change of magnitude of the birefringence. The rotation angle θ can be approximated by:

$$\theta = \Delta n_a / 2\Delta n_i \qquad (4.8)$$

Light propagating along the fibre and originally polarized along the *y*-axis will be decomposed into components polarized along the prime axes when entering a compressed region (see Figure 4.6). The phase of the light in the two polarizations will change by π radians in half a beat length.

After half a beat length, the force is removed and the light decomposed back into components along the original axis with relative power $\cos^2(2\theta)$ in the *y*-polarization and $\sin^2(2\theta)$ in the *x*-polarization. This phenomenon of accumulated power transfer has been achieved via a set of periodic pressure regions. Using Jones calculus, a Jones matrix T has been calculated for a single $L_b/2$ length stressed region and $L_b/2$ unstressed region.

Repeating N times this structure, complete power coupling from one polarization to the other can be achieved by applying a force F to the N ridges so that $2N\theta = \pi/2$. The optimal force deduced from equations (4.7) and (4.8) is:

$$F = (\pi \ \Delta n_i \ 2b)/N \ S \ n^3 \ K \qquad (4.9)$$

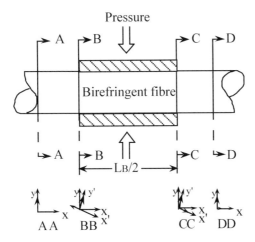

Figure 4.6. Scheme of a coupler using birefringent fibre. The light, linearly polarized along the y-axis in region A, enters the stressed region at location B and decomposes into components along the stressed polarization axes. At location C after half a beat length, a maximum amount of light has been transferred to the x polarization axis (location D) [After Youngquist et al. 1985].

A coupling ratio greater than 32 dB can be achieved with a force of about 220 g with 10 ridges at an optical wavelength of 633 nm (Youngquist et al. 1985). So the two principal polarizations can be coupled by subjecting the fibre to a stress that is spatially periodic along the length of the fibre.

Using an acoustic wave the stress may be varied in time. The acoustic wave varies the pressure about that point causing a modulation of the power present in the two polarizations. The optical power can be transferred between polarizations dynamically. Optimum coupling occurs when stress is applied at 45° to the principal axes. The amount of power $P_2(L)$ coupled after a length $L = NL_b$ is given by

$$P_2(L) = P_1(0) \ \sin^2(CNL_B) \tag{4.10}$$

where $P_1(0)$ is the initial power in one polarization, N is the number of ridges, C the coupling coefficient and L_b the beat length. Assuming that the acoustic wave causes a peak variation of δC in the coupling coefficient C, the modulation depth M, defined as the peak-to-peak variation of the output optical power expressed as a fraction of the total power around $CL = \pi/4$, is given by:

$$M = \sin(2 \ L \ \delta C) \tag{4.11}$$

Static pressure is applied to the ridges until half of the power initially in one polarization is coupled to the other ($CL = \pi/4$). An example is given by Risk and Kino (1986) using an elliptical core birefringent fibre. The surface acoustic waves are generated on a fused quartz substrate block using edge-bonded lead zirconate titanate (PZT) transducers. When an output polarizer is used to pass only one of the principal axis polarizations, the acoustically induced exchange of power between polarizations is seen as intensity modulation.

4.4 POLARIZATION DEVICES

4.4.1 Polarizers

Single-mode fibre type polarizers are important; several principles are used for creating such devices for optical fibre communications and sensors systems (Table 4.1).

Fibre-optics polarizers have been fabricated by using a birefringent crystal placed close to the fibre core, or by depositing a metal film.

Table 4.1 Polarizer made from single-mode fibres

Principle	Authors
Birefringence properties of nematic liquid evanescent field	Liu *et al.* 1986
Birefringent crystal	Bergh *et al.* 1980
Metal clad = attenuation difference between TM_o and TE_o modes	Feth and Chang 1986
Cut off principle (silver)	Hosaka *et al.* 1983
(aluminium)	Dyott *et al.* 1987
(Indium)	Zervas 1990
(Nickel)	
Metal clad. Theoretical analysis	
Circular core	Yu and Wu 1988
Elliptical core	Pilevar *et al.* 1991
2 Metal clad + dielectric superstrate	Zervas and Giles 1989
1 Metal clad + dielectric superstrate	Pilevar *et al.* 1991
	Johnstone *et al.* 1990
Resonant excitation of surface wave	Dragila and Vakovic 1991
Circular polarizer composed of a metal coated fibre and $\lambda/4$ platelet fabricated on a birefringent fibre	Hosaka *et al.* 1983

4.4.1.1 Linear polarizer using birefringent material

In one technique the evanescent field of the guided light interacts with a birefringent crystal, causing light of undesired polarization to couple out of the fibre (Bergh et al. 1980). Part of the cladding is removed over a small length of fibre to allow access to the evanescent field. The birefringent crystal used to replace the removed position of cladding is a potassium pentaborate crystal ($KB_5O_8\text{-}4H_2O$) with three refractive indices which we may call n_a, n_b and n_c. If the effective refractive index of the crystal (see Figure 4.7) is greater than the effective index of the waveguide, the guided wave excites a bulk wave in the crystal and light escapes from the fibre. If the effective refractive index of the crystal is less than the effective index of the waveguide no bulk wave is excited and no light escapes from the fibre. The crystal index for the polarization perpendicular to the crystal fibre interface is n_b and the crystal is cleaved perpendicular to the b axis. The crystal index for the orthogonal polarization $n_\perp$ lies between n_c and n_a following the equation

$$n_\perp = \left(\frac{\sin^2\theta}{n_c^2} + \frac{\cos^2\theta}{n_a^2}\right)^{-1/2} \tag{4.12}$$

where θ is the angle between the polarization and the a axis of the crystal. The extinction ratio is defined as:

$$S' = 10\,\log_{10}(P_{\text{pe}}/P_{\text{pa}}) \tag{4.13}$$

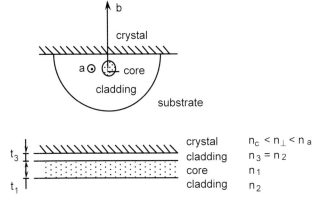

Figure 4.7. Scheme of polarizers using birefringent crystal: (a) Crystal b-axis is perpendicular to crystal-cladding interface, a and c are in plane of interface, (b) four-layer planar model [After Bergh et al. 1980].

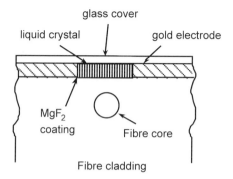

Figure 4.8. Cross section through coupler-half showing positional relationship between liquid crystals, electrodes and fibre core [After Liu et al. 1988].

where P_{pa} and P_{pe} are the power of the two modes corresponding to the parallel and perpendicular polarization.

For example at $\lambda = 633$ nm, $n_a = 1.49$ and $n_b = 1.43$ and $n_c = 1.42$, extinction ratios of 60 dB and low losses for the desired polarization have been obtained.

Another possibility for an in-line polarizer is to use the birefringent properties of a nematic liquid crystal (Liu et al. 1986). The device is a coupler-half (see Figure 4.8). When nematic alignment is produced in the liquid crystal, one polarization state will see the high index and will not be guided into the fibre while the orthogonal polarization will see the low index and will remain guided. To reach the nematic phase the crystals must be heated to between 62° and 85°C. This device is proposed for applications in which a high-speed response is not required. The amplitude modulation has been demonstrated by controlling the alignment of the liquid-crystal molecules with an electric field for a frequency limited to a few kilohertz and a polarization discrimination of 45 dB.

4.4.1.2 Linear polarizer using thin metal films

A second type of device uses a thin metal film (silver or gold). Metal-clad optical fibre polarizers are based on either the differential attenuation of the two polarization modes or the cut-off of the TE_{00} mode (Eickhoff 1980, Feth and Chang 1986, Hosaka et al. 1983a, Dyott et al. 1987, Zervas 1990, Johnstone et al. 1990, Kutsaenko et al. 1994). The core diameter of a single-mode fibre is reduced by grinding and polishing so that the fibre is below the cut-off condition in the core. A thin metal film is deposited upon the top of the polished fibre (see

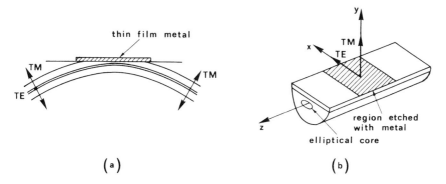

Figure 4.9. Cross section of a metal clad fibre optic polarizer: (a) in silica block, (b) D-shaped fibre.

Figure 4.9a). Only the TM wave can satisfy the boundary condition and propagates along the interface. Light of TM polarization propagating in the fibre is converted into a surface plasmon wave propagating through the interaction region, and is converted back into a TM polarization light wave in the fibre. The TE mode is radiated into the cladding when it reaches the interaction region. The TE mode can be coupled to another waveguide upon the thin metal layer in order to be guided without losses along this planar waveguide. The TM mode is very sensitive to the refractive index of the planar waveguide superstrate.

It is also possible to use silver and gold deposited onto a flat surface of an optical quality fused-silica superstrate. The superstrate is placed on the polished surface and index matching oil is allowed to flow between the two by capillarity. Results range from an extinction ratio of 50 dB and insertion loss of 6 dB to an extinction ratio of 45 dB and insertion loss of 1 dB.

Another method of fabrication has been proposed by Hosaka *et al.* (1983b). The single-mode fibre has a concentric core and a silica cladding having a B_2O_3 doped silica portion (Figure 4.10). The cladding is etched off asymmetrically by using the etching-speed difference between pure silica and doped silica (49% H.F.). On to the etched part an aluminium film is evaporated. A 37 dB maximum extinction ratio is obtained when the polarizer is 40 mm long (Figure 4.11).

In another example of this type of device, a metal film polarizer is deposited on an etched D-shaped fibre which has a birefringence induced by an elliptical core (see Figure 4.9b) (Dyott *et al.* 1987, Yu and Wu 1988). The suppressed mode is the $_eHE_{11}$ mode with the electric field along the minor axis of the elliptical guide and normal to the flat of the D fibre. Assuming that the coupling is incoherent, and is

144   *Devices using polarized light*

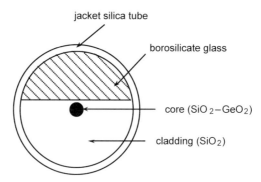

Figure 4.10. Preform structure from which the fibre type polarizer was drawn [After Hosaka *et al.* 1983].

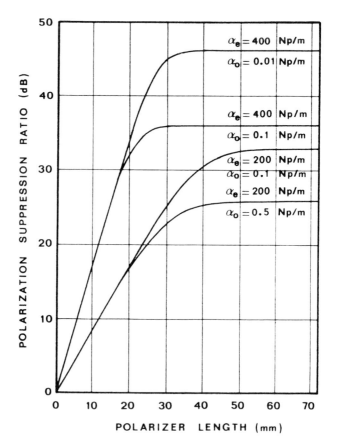

Figure 4.11. Theoretical variation of suppression ratio S' versus polarizer length for various of α_e and α_o [After Dyott *et al.* 1987].

constant along the length L, the mode power reaching a distance z is given by:

$$P_e(z) = P_i \exp(-\alpha_e z)$$

for the $_eHE_{11}$ mode and

$$P_o(z) = P_i \exp(-\alpha_o z) \qquad (4.14)$$

for the $_oHE_{11}$ mode. α_e and α_o are the attenuations in nepers per meter and P_i the input power launched equally into each mode at $z=0$.

In an element dz, the amount of power P_o transferred to P_e is attenuated by the polarizer along $L - z$. Integrating over the length L, the total emerging transferred power is P_t. So the total even mode power P_E at L is $P_E = P_t + P_e(L)$. Assuming that $\alpha_e \gg \alpha_o$, the polarization suppression ratio S could be expressed as:

$$S = P_o(L)/P_E = \{\exp(-\alpha_e L) + (\alpha_o/\alpha_e) \exp(\alpha_o L)[1 - \exp(-\alpha_e L)]\}^{-1} \qquad (4.15)$$

The theoretical variation of S with the polarizer length is given Figure 4.11. The critical length L_c at the value where saturation occurs is deduced when $P_E / P_o(L)$ is near 0.

$$L_c = (1/\alpha_e) \ln(\alpha_e/\alpha_o) \qquad (4.16)$$

and the saturated polarization suppression ratio is $S_s = \alpha_e / \alpha_o$ which tends to a high value if $\alpha_e \gg \alpha_o$.

Using equation (4.13) with P_{pa} and P_{pe} expressed as equation (4.14) the extinction ratio becomes:

$$S'(dB) = 10 (\alpha_e - \alpha_o)L/\ln 10 \qquad (4.17)$$

S' is a linear function of the polarizer length (see Figure 4.12), α_o being the attenuation coefficient for a mode polarized in the plane of the metal–fibre interface (TE), α_e the attenuation coefficient for a mode polarized perpendicularly (TM). S' corresponds to the polarization suppression ratio expressed in dB ($S' = 10 \log_{10} S$).

A rectangular core waveguide model acquires the polarization characteristics of a metal-clad elliptical - core fibre polarizer (Pilevar et al. 1991a,b). The losses of the TM-like and TE-like models of such a polarizer have been obtained as a function of depth of polishing. The attenuation for the TM-like polarization is very large compared to that of the TE-like polarization. Such a metal-clad structure behaves as a TE-pass polarizer. A peak in the attenuation of the TM-like mode is also observed (see Figure 4.13) for a given depth of polishing.

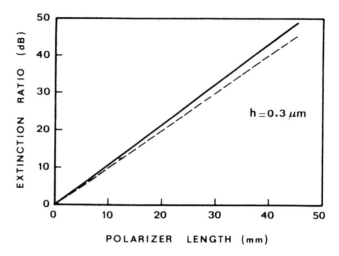

Figure 4.12. Relation between extinction ratio S' and polarizer length L. Solid line, theoretical prediction; broken line experimental data (Hosaka *et al.* 1982) [After Yu and Wu 1988].

4.4.1.3 Linear polarizer using multiple interfaces

Another structure consisting of two metallic layers, one of aluminium and one of chromium has been reported (Zervas and Giles 1989, Pilevar *et al.* 1991a) (see Figure 4.14a). The addition of chromium results in an increased polarization extinction ratio over a wide range of superstrate refractive indices. Using a planar waveguide model Pilevar *et al.* (1991a,b) studied the variation of the TM mode loss with depth of polishing h and with the superstrate refractive index n_1 (see Figure 4.14b). The increase of the polarization extinction ratio is due to the metallic behaviour of the chromium, which supports a surface plasmon wave.

High quality optical fibre polarizers have also been made by depositing thin metal films on polished fibres, but with a dielectric superstrate. For example, thin nickel films are deposited on a polished fibre block and coated with a dielectric (Zervas 1990, Zervas and Giles 1989, Johnstone *et al.* 1990) (see Figure 4.15). Surface plasmon polaritons at optical frequencies can be supported by the thin metal film sandwiched between two dielectrics. These surface (TM) waves are guided by single or multiple metal/dielectric interfaces. The polarization extinction is achieved by efficiently exciting TM-polarized surface plasmon polariton waves, supported by the metal film, by evanescent field interaction with the fundamental fibre mode. TM-like polarized

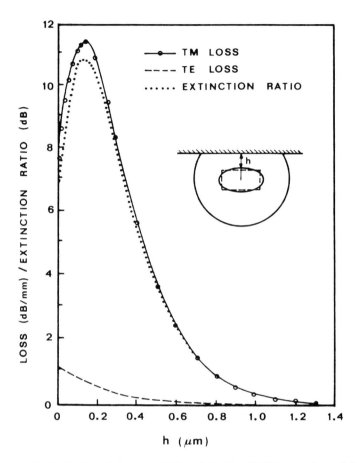

Figure 4.13. Variation of the loss (dB/mm) for the TM-like mode (solid curve) and TE-like mode (dashed curve) and the extinction ratio (dotted line) of the polarizer as a function of the depth of polishing [After Pilevar et al. 1991].

light is coupled and extinguished when phase matching conditions are achieved and the TE-like polarization passes through with minimal losses. Polarization extinction ratios in excess of 55 dB have been achieved with an insertion loss of less than 1 dB.

The conditions for coupling between the fibre mode and one of the surface plasmon modes of a thin metal film, are (Johnstone et al. 1990):

$$k_0 \, t(n_e^2 - n_2^2)^{1/2} = \tan h^{-1}(A_1) + \tan h^{-1}(A_3) \tag{4.18}$$

where $k_0 = 2\pi/\lambda$, n_e is the effective index $n_e = \beta/k_0$, β is the propagation constant of the plasmon mode, t the metal thickness. A_1 and A_3 are

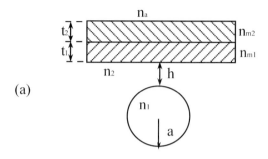

(a)

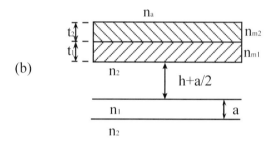

(b)

Figure 4.14. Scheme of a dual metal coated in-line fibre optic polarizer: (a) dual metal-clad fibre polarizer structure, (b) equivalent planar waveguide structure [After Pilevar et al. 1991a].

given by

$$A_i = -\left(\frac{n_2}{n_i}\right)^2 (n_e^2 - n_i^2)^{1/2}/(n_e^2 - n_2^2)^{1/2} \quad (i = 1, 3) \qquad (4.19)$$

where n_2, n_1 and n_3 are the refractive indexes respectively of the metal and of the two dielectrics. Equation 4.18 has four significant solutions corresponding to the bound and leaky symmetric modes s_b and s_l and the bound and leaky antisymmetric modes a_b and a_l (see Figure 4.16).

Using equation 4.18, the effective index of each of the four plasmon modes of a thin metal film can be determined as a function of film thickness for several values of the overlay index n_3. The antisymmetric plasmon modes do not play a role in device operation. Coupling occurs to the bound symmetric mode over a continuous range of combination of overlay index n_3 and metal thickness for which the plasmon effective

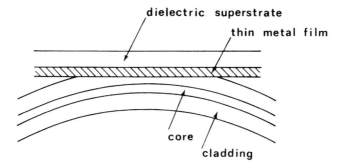

Figure 4.15. Schematic diagram of the thin metal film plasmon polarizing structure.

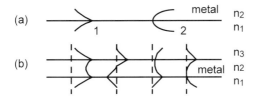

Figure 4.16. General forms of the electric field distribution of surface plasmon supported by a thin metal film: (a) metal dielectric interface, (b) thin metal film, s_b, s_l are bound and leaky symmetric modes, a_b, a_l are bound and leaky antisymmetric modes [After Johnstone et al. 1990].

index n_e equals the fibre effective index n_{ef}. Efficient fibre mode to plasmon mode coupling and high polarization extinction ratio is obtained if $n_e = n_{ef}$. Several metal such as aluminium, chromium, and silver can be chosen for practical polarizers. By optimizing the design parameter, high quality could be achieved with extinction ratios in excess of 50 dB for 1.3 μm devices operating in the leaky mode and 60 dB for 0.632 μm devices operating in the bound mode. The loss for the transmitted polarization *TE* is less than 0.5 dB.

Another in-line fibre-optics polarizer is based on resonant excitation of surface waves propagating along a thin metal implanted in a fibre cladding (Dragila and Vukovic 1991). Such a device allows for resonant excitation of the so-called surface mode only by an even LP_{01} mode, and simultaneously for its absorption within the metal film. The metallic film is very thin ($\approx \lambda$), located far from the core (4–10 λ) (see Figure 4.17). The polarization that is not involved in excitation of surface wave (the odd LP_{01} mode) is only slightly perturbed, in contrast to the other methods.

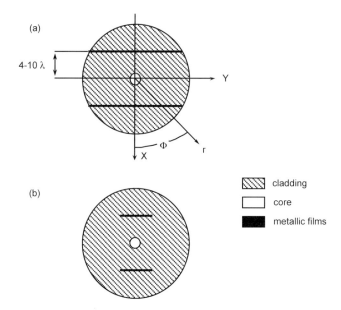

Figure 4.17. Schematic diagram of a cross section of an optical fibre polarizer with metallic films inserted into the cladding: (a) transverse electric polarized predominantly along X and magnetic along Y for one state of polarization (r and φ are the polar coordinates), (b) another type of polarizer [After Dragila and Vukovic 1990].

4.4.1.4 Circular polarizers

A fibre circular polarizer composed of a metal-coated fibre and a $\lambda/4$ platelet made by cutting a birefringent fibre to an appropriate length can be manufactured (Figure 4.18).

Linearly polarized light which passed through the fibre polarizer is launched into the following birefringent fibre of beat length L_b with polarization angle $\theta = \pi/4$ incident to the birefringent fibre principal axes. The fibre polarizer ($\cong 5$ cm in length) has been fabricated on a section of the birefringent fibre by evaporating Aluminium of 1 000 Å thickness. The length L of the birefringent fibre part following the polarizer has been set at $L = [2N + 1]\, L_b\, /\, 4$ (N is an integer) to operate as a $\lambda/4$ platelet. When left -and right- circularly polarized light was launched into the device, a 17.6-dB power ratio was obtained within an angular deviation of $|\Delta \theta| \leq 1°$ from the optimum angle between the fibre polarizer axis and the major axis of the birefringent fibre. When used with a light source, such as a laser diode, this device will operate as a quasi-isolator because the light reflected from the output end (the main factor backing the light source), can be eliminated.

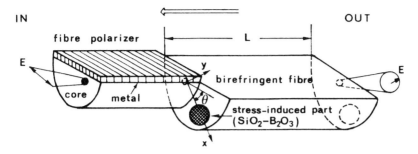

Figure 4.18. Scheme of a fibre circular polarizer [After Hosaka et al. 1983].

4.4.2 Depolarizers

As seen at the beginning of this chapter, when polarized light is launched in a single-mode fibre, the polarization is maintained along a large length of the fibre but the state of polarization (SOP) fluctuates randomly due to external influences such as temperature or bends. When a polarizer is inserted, SOP fluctuation gives rise to intensity noise. Depolarizing devices have been proposed to reduce this kind of noise (Takada et al. 1986, Böhm et al. 1983, Hillerich and Weidel 1983). A statistical approach based on probability density function of optical polarization states can be applied to these depolarizing devices (Van Deventer 1994).

A fibre Lyot depolarizer converts different spectral components of polarized input light into different polarization states at the output, so that the output light appears unpolarized if averaged over the spectrum. An example has been realized by using a birefringent fibre, which is cut and then spliced again after turning the principal axis of one end by an angle of 45° (Böhm et al. 1983). The single-mode fibre used is a fibre with a large linear birefringence and a beat length $L_b = 2\pi / \beta = 4$ mm at 633 nm. For the fabrication of the Lyot depolarizer two fibres pieces with lengths of 5 m and 10 m are used. The performance of the device depends critically on the angle between the principal axes of the two fibres. A residual polarization of about 1% has been obtained with a superluminescent diode.

Another depolarizer system is consisting in a Mach-Zehnder interferometer. The light passing through the half wave plate HW1 is divided into P and S waves by the polarization beam splitter BS1 (see Figure 4.19) (Takada et al. 1986). The S wave is launched into a polarization maintaining PANDA fibre where one principal axis is parallel to the polarization of the S wave. Output light after passing through the half wave plate HW2 and the polarizer PO1 is then coupled orthogonally with the P wave at the beam splitter BS2. The fibre is a delay line of the S wave, achieving a large group delay time difference of 5 ns/m. When the group delay time difference is longer than the coherence time of the light the two orthogonal components of the electrical fields are completely uncorrelated. Adjusting the rotation angles of the two half wave plates HW1 and HW2, the P and S waves are superimposed with equal power at the second beam splitter BS2. Light having a 100 MHz spectral width can be depolarized with high coupling efficiency by using a polarization maintaining fibre delay line of a few meters in length. Light having a 51 MHz spectral width can be depolarized to within 14 dB using only 2 m fibre length.

4.4.3 Polarization state controllers

In coherent optical fibre communications and interferometric optical fibre sensor devices, the sensitivity of the systems is dependent on the matching of the state of polarization (SOP) of the recombining beams. In coherent optical transmittance systems it is necessary to match the time varying SOP of the incoming signal with the local oscillator one. It is important that matching is maintained continuously, irrespective of variations of the signal SOP, as momentary mismatch can result in unacceptable data loss. The state of polarization of signals in installed fibres varies slowly enough to permit polarization compensation. A practical SOP controlling system must change the polarization endlessly to prevent momentary signal loss during the reset of any constituent components.

Several schemes of SOP control devices have been proposed (see Table 4.2). Several kinds of polarization control schemes have been published using the elasto-optic properties of silica by introducing: controlled squeezing (Ulrich 1979, Johnson 1979, Granestrand and Thylen 1984, Honmou *et al.* 1986), bending of the fibre (Lefebvre 1980) piezoelectric cylinders wound with polarization maintaining fibres

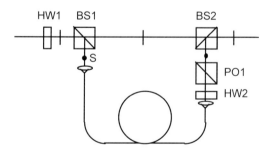

Figure 4.19. Set up for single-mode fibre optic depolarizer [After Takada *et al.* 1986].

(Walker and Walker 1990), two short sections of linearly birefringent fibres (Tatam *et al.* 1987, Pannel *et al.* 1988), nematic liquid crystals (Rumbaugh *et al.* 1990, Barnes 1988), or magneto-optics properties like Faraday rotation (Okoshi *et al.* 1985). A polarization state controller should be able to convert any incident polarization state to any other state at the output. However, in practical applications the requirement is to convert a fixed linear state to an arbitrary state or vice-versa.

4.4.3.1 Polarization controller using stressed fibres

In a bent fibre, the material in the central region containing the core receives a stress in the direction xx' of the curvature radius R (see Figure 4.20).

This stress is the source of the uniaxial negative-induced birefringence. Two principal axes are considered: a fast extraordinary axis $x'x$ and a slow ordinary axis $y'y$. The changes Δn_i ($i = x$ or y) in the index n_1 of the core (initially isotropic) are given by:

$$\Delta n_x = (n_1^3/4)(p_{11} - 2\sigma p_{12})(b/R)^2$$
$$\Delta n_y = (n_1^3/4)(p_{12} - \sigma p_{12} - \sigma p_{11})(b/R)^2 \quad (4.20)$$

where p_{ij} are terms of the photoelastic tensor, σ is Poisson's coefficient, b the outer radius of the fibre. For silica $\sigma = 0.16$, $p_{11} = 0.121$, $p_{12} = 0.270$, $n_1 = 1.46$ for $\lambda = 0.633$ µm, thus:

$$\Delta n_x = n_e - n_1 = 0.027(b/R)^2$$

$$\Delta n_y = n_o - n_1 = 0.160(b/R)^2$$

and

$$\delta n = n_e - n_o = -d(b/R)^2 = -0.133(b/R)^2 \quad (4.21)$$

where $d = 0.133$.

This birefringence integrated along the fibre can give, between two modes, a total phase delay of π, $\pi/2$ or $\pi/4$ if it is possible to coil a length L of fibre with a calculated number of turns N.

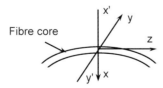

Figure 4.20. Scheme of a bent fibre.

Table 4.2 Polarization control schemes [After Pannel et al. 1988]

Type of SOP control scheme	Insertion loss	Temporal response	Mechanical fatigue	References
1. *Mechanical: fibre optic*				
Fibre squeezers	Low	Medium	Yes	Youngquist et al. 1985
Fibre rotators	Low	Slow	Yes	Uttam and Culshaw 1986
Rotatable fibre coils				
(i)	Low	Slow	Yes	Kim et al. 1986
(ii)	Low	Slow	Yes	Imai et al. 1985
				Ulrich 1979
Rotatable fibre cranks	Low	Slow	Yes	Johnson 1979
Linearly birefringent fibre	Low	Slow	Yes	Pannel et al. 1988
2. *Electro and magneto-optic*				
Electro-optic crystals	High	Fast	No	Ulrich and Johnson 1979
Faraday rotators	Low	Fast	No	Lefevre 1980
				Granestrand and Thylen 1984
3. *Mechanical: conventional optics*				
Rotatable phase plates	Medium	Slow	No	Matsumoto and Kano 1986
Non-rotating linear polarizer mask (passive technique)	Low	Fast	No	Okoyshi et al. 1985

The phase shift is then given by:

$$\phi = 2\pi(\delta n\, L/\lambda) = 2\pi/m$$

where $m = 2, 4, 8$. Using equation 4.21 and $L = 2\pi R N$ we have

$$R = 2\pi d\, b^2\, Nm/\lambda \qquad (4.22)$$

These devices, which are equivalent to phase delay plates, have been described by Lefevre (1980), and by Ulrich and Simon (1979). For coil diameters below 3–4 cm, the polarization holding degrades very quickly with a decreasing diameter. The cross coupling is due to a microbending effect (Rashleigh and Marrone 1983).

A polarization controller uses twists in the coils. Let us consider a coil fixed at the points B and C (see Figure 4.21). A rotation of the plane of the coil through an angle α induces opposite twist in sections BA and AC. The twist in BA rotates the incident polarization through an angle $\alpha' = t\alpha$, between A and B. The angle between the polarization in B and the principal axis of the coil hence has a variation of $(1 - t)\alpha$ and the effect of the twist is small in comparison with the effect of the direct rotation of the principal axes of the coil. It is the same thing for the twist between A and C. A device using these effects consists of two $\lambda/4$ coils to control the ellipticity and one $\lambda/2$ to control the orientation of any desired output polarization (Lefevre 1980). With an analyser, a polarizer and several coils a tunable wavelength filter device has been achieved (Gao et al. 1993).

A fibre optic phase modulator (see Chapter 3 § 3.8.1) comprising two coils of single-mode fibres wound on cylindrical piezoelectric elements can reduce the modulation retardation by two orders of magnitude less than for a single coil of similar phase modulation amplitude. This is achieved by adjusting the angle between the axes of the cylinders. This type of device is used in optical fibre interferometers (Luke and al 1995).

A polarization maintaining fibre has been proposed by Walker and Walker (1990). It comprises four piezoelectric cylinders wound with 85 turns of *PM* fibre, which are spliced together with the principal axes of the *PM* fibre mutually aligned at $\pi/4$ (see Figure 4.22). The fibre is highly birefringent, with a beat length of a few millimetres. Application of a voltage causes the piezoelectric cylinders to expand slightly and to stretch the fibre and thus modify the birefringence. The result is a controlled retardation.

A device which offers a simple way to adjust any state of polarization in a single-mode fibre has been described by Krath and Scholl (1991). The polarization transformer consists of a rotatable fibre squeezer, a standard single-mode fibre, and two blocks holding the fibre in position (see Figure 4.23). The fibre is glued to the outer blocks

156   *Devices using polarized light*

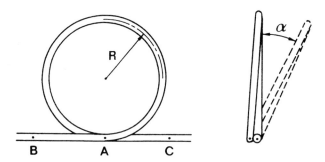

Figure 4.21. Fibre loop equivalent to phase delay plate [After Lefevre 1980].

as well as to the squeezer, which rotates around the fibre axis. There are two fibre parts of length L which can be twisted around one another. A lateral force can be applied to the middle part of the fibre by means of the squeezer. The polarization at the end can be set as desired using a combination of adjustments i.e., the rotation α of the squeezer and the pressure F applied to the fibre. The length over which the lateral force is applied is not critical.

4.4.3.2 Polarization controller using intrinsic fibre birefringence

Another all-fibre polarization state controller consists of two short sections of linearly birefringent fibre oriented with their eigenaxes at 45° in respect to each other. This has been constructed by Tatam *et al.* (1987, 1988) and by Pannell *et al.* (1988) (see Figure 4.24). A linearly polarized light beam is launched into the first fibre.

Both eigenmodes are equally populated. The evolution of the state of polarization is a function of axial strain. Axial strain applied to both fibre elements is used to control the propagation constant. The output is coupled into the second element which is orientated with its eigenmode axes at 45° with respect to the first fibre. Controlling the output from the first fibre allows variable population of the eigenmodes of the second fibre. The range varies from the equal population of both eigenmodes to the population of only one mode, resulting in a linear output state.

Using the Jones calculus the transmitted electrical field can be determined for each polarized input state. As an example for a horizontally polarized incident electrical field the transmitted field $\boldsymbol{E_t}$ is:

$$\boldsymbol{E_t} \cong \exp(i\,\Delta\phi_1/2) \begin{bmatrix} \cos(\Delta\phi_1/2) \\ \exp(i\delta)\,\sin(\Delta\phi_1/2) \end{bmatrix} \qquad (4.23)$$

Polarization devices 157

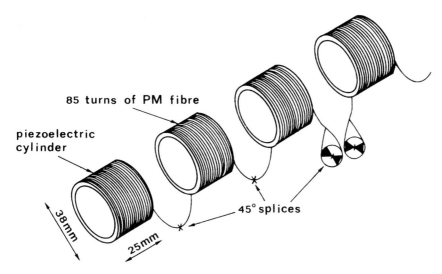

Figure 4.22. Polarization-maintaining fibre controller [After Walker and Walker 1990].

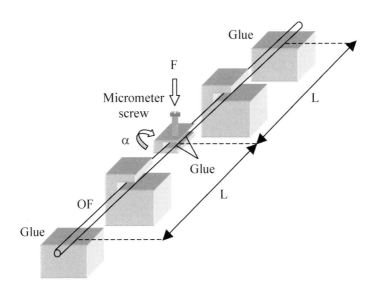

Figure 4.23. Schematic of the squeezer.

where $\delta = \Delta\phi_2 - \pi/2$. $\Delta\phi_1$ represents the polarization azimuth and $\Delta\phi_2$ the modal retardance between the orthogonal linear components, corresponding to the polarization ellipticity. Such a device converts a horizontal or vertical linear state to any other.

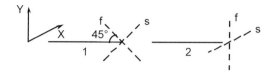

Figure 4.24. Scheme of a SOP controller made with linearly birefringent fibres 1 and 2. f: fast, s: slow [After Tatam et al. 1987].

For the control and modulation of the azimuth of a linearly polarized beam, an all-fibre device uses a Mach-Zehnder interferometer with the two arms made of birefringent fibres (see Figure 4.25).

It produces orthogonal circular states of polarization which recombine in the final directional coupler. Azimuth control is then achieved by relative phase modulation using a piezoelectric element around which fibre is wound (Tatam et al. 1988). Two linearly polarized beams S_1 and S_2 are produced by amplitude division and are incident upon quarter-wave plates, thus producing two beams with orthogonal circularly polarized states that are recombined on the final beamsplitter. The beamsplitters are single-mode fibre optic couplers and wave plates are made from optical fibres.

4.4.3.3 Polarization controller using liquid crystals

Liquid crystals are also used to change the state of light polarization and can be applied to a continuous control system for coherent detection. Using nematic liquid crystals with an antiparallel surface alignment, liquid crystal acts as a retarder in a polarization controlling system (Rumbaugh et al. 1990). The retardance Γ is a function of molecular alignment which is controlled by an applied electric field. It is given by the equation:

$$\Gamma = (2\,\pi\,L/\lambda)\,[n_e(V) - n_o] \qquad (4.24)$$

where L is the cell thickness, and n_e (V) and n_o are the indices of refraction for the extraordinary and ordinary rays respectively. The extraordinary index is a function of applied voltage V.

An optical fibre polarization switch based on liquid crystal technology has been demonstrated by Barnes (1988). Switching voltages of about 3V has been used and switching speeds of a few Hz obtained. The device may also be used as a polarization controller in a coherent optical scheme.

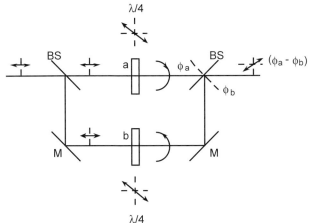

Figure 4.25. Concept of a device producing a linearly polarized state with controlled azimuth: BS beam splitter, M Mirror, $\lambda/4$ waveplate [After Tatam et al. 1988].

4.4.4 Isolators

Isolators are components able to prevent optical feedback, and are useful for isolation of light sources. Optical feedback can modify the stability of the diode causing *partition noise*. Isolators are fabricated in a large variety of magneto-optical materials such as garnet crystals (yttrium iron garnet (YIG), terbium aluminium garnet, etc.), glasses and crystals (Chang and Sorin 1990). These components act as Faraday rotators with a non-reciprocal rotation for direction of polarization in the NIR range. It is possible to obtain a high performance in polarization-independent isolators insensitive to temperature and wavelength by an arrangement of YIG crystals used as Faraday rotators.

An isolator can also be fabricated using plastics optical fibre surrounded by a series of permanent magnets (Muto *et al.* 1991). Another device working as a quasi-isolator using a fibre polarizer has been described in § 4.4.1.4 (Hosaka *et al.* 1983). An in-line optical isolator without lenses can be constructed by embedding an isolator chip made by two pairs of rutile wedges and garnet plates in a thermally expanded core fibre (Sato *et al.* 1999).

4.5 POLARIMETERS – INTERFEROMETERS

The phase of a light wave propagating in an optical fibre is more sensitive to external influences than any other propagation parameter. Two approaches have appeared in the development of sensitive fibre optic interferometric sensors. The optical fibre Mach-Zehnder interferometer uses two single-mode fibres to detect the relative phase shift of light propagating in the two arms of the interferometer (cf. Chapter 3). Alternatively, single fibre interferometers called polarimeters rely on the relative phase displacement of two polarization eigenmodes (see Figure 4.26). Performances of fibre optic polarimeters and applications to sensors are analysed in several sources books (Mermelstein 1986, Rogers 1985, Culshaw 1983, Donati et al. 1988, Youngquist et al. 1985, Dakin and Wade 1984, Rashleigh and Marrone 1983, Azzam 1990). The phase delay ϕ between the two polarization eigenmodes of a high birefringence single-mode fibre is

$$\phi = k_0 \, BL$$

where B is the fibre birefringence (see equation 1.30), L the propagation length and $k_0 = 2\pi/\lambda_0$. For example, an axial strain in a high birefringence fibre produces a modification in the birefringence and introduces a strain-induced phase delay.

The design of a polarimeter is illustrated in Figure 4.27. It consists of two lengths of optical fibres L_1 and L_2. For example, length L_1 is wrapped around a piezoelectric (PZT) cylinder, and provides a voltage-controlled phase delay for maintaining quadrature. Length L_2 is the sensor length. The principal axes of the two fibres are rotated of 90° with respect to one another. Fast and slow axes are aligned, providing optical common mode rejection (Dakin and Wade 1984). Allowance has been made for a small deviation θ.

Assuming that the two polarization eigenmodes are uncoupled and that the light is monochromatic, linearly polarized light of an amplitude E_0 is launched in fibre 1 at an angle of 45° to the principal axes. At the end of fibre 1 the field components along the principal axes are:

$$E_x(L_1) = \left(E_0/\sqrt{2}\right) \exp(ik_0 \, n_x \, L_1)$$
$$E_y(L_1) = \left(E_0/\sqrt{2}\right) \exp(ik_0 \, n_y \, L_1)$$

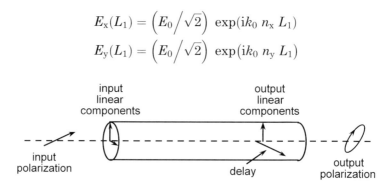

Figure 4.26. Linear polarimetric device equivalent to Mach Zehnder

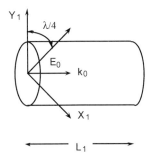
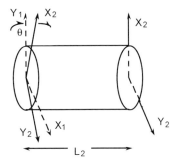

Figure 4.27. Design of polarimeter sensor with optical common mode rejection. Fibre segment L_2 is rotated by approximately 90° with respect to fibre segment L_1. Angular misalignment is denoted by θ [After Dakin and Wade 1984].

Onto the principal axes of fibre 2 the output electric field components are:

$$\begin{cases} E_x = \exp(ik_0\, n_y\, L_2)\left[E_x(L_1)\cos\theta + E_y(L_1)\sin\theta\right] \\ E_y = \exp(ik_0\, n_x\, L_2)\left[-E_x(L_1)\sin\theta + E_y(L_1)\cos\theta\right] \end{cases}$$

E_x and E_y are projected onto the transmittance axes of a prism polarizer. Mermelstein (1986) constructed a polarimeter from 50 m of highly birefringent bow-tie fibre. Minimum detectable rms phase delays below 1.0 μrad with a 1 Hz bandwidth are achievable at frequencies greater than 50 kHz. Several other applications for current and magnetic field measurements are presented by Donati et al. (1988), Rogers (1985) Laming and Payne (1989). Light can also be coupled between the two principal polarizations of birefringent fibre by using a travelling acoustic wave producing a spatially periodic stress in the fibre. By changing the acoustic frequency the wavelength at which coupling occurs can be tuned (Risk et al. 1986).

Using a set of PM fibre pieces separated by several fibre optics polarizers, a wavelength filter with a free spectral range of a few nanometres and a full width half maximum of smaller than 1nm can be achieved, operating as a polarization interference filter (Reichel and al 1993).

REFERENCES FOR CHAPTER 4

Azzam RMA 1990 *IEEE Photonics Technol. Lett.* **2** 893–5
Barnes WL 1988 *Electron. Lett.* **24** 1427–9

Bergh RA, Lefevre HC and Shaw HJ 1980 *Opt. Lett.* **5** 479–81
Böhm K, Petermann K and Weidel E 1983 *IEEE J. Light wave Technol.* **1** 71–4
Bricheno T and Baker V 1985 *Electron. Lett.* **21** 251–3
Cancellieri G, Chiaraluce F and Gianfagna A 1990 *Opt. Comm.* **78** 230–6
Chang KW and Sorin WV 1990 *Optics Lett.* **15** 449–51
Dakin JP and Wade CA 1984 *Electron. Lett.* **20** 51–3
Donati S, Annovazzi-Lodi V and Tambosso T 1988 *IEE Proc. PtJ.* **135** 372–83
Dragila R and Vukovic S 1991 *Intern. J. of Optoelectron.* **6** 35–45
Dyott RB, Bello J and Handerek V A 1987 *Opt. Lett.* **12** 287–9
Eickhoff W 1980 *Electron. Lett.* **16** 762–3
Feth JR and Chang C L 1986 *Opt. Lett.* **11** 386–8
Fujii Y 1991 *J. Light wave Technol.* **9** 456–60
Gao P, Bassi P and Zoboli M 1993 *J. Opt. Commun.* **14** 128–33
Granestrand P and Thylen L 1984 *Electron. Lett.* **20** 365–6
Hakli B W 1996 *J. Light wave Technol.* **14** 2202–8
Hillerich B and Weidel E 1983 *Opt. Quantum Electron.* **15** 281–7
Honmou H, Yamazaki S, Emura K, Ishikawa R, Mito I, Shikada M and Minemura K 1986 *Electron. Lett.* **22** 1181–2
Hosaka T, Okamoto K and Noda J 1982 *IEEE J. Quantum Electron.* QE **18** 1569–72
Hosaka T, Okamoto K and Edahiro T 1983a *Opt. Lett.* **8** 124–6
Hosaka T, Okamoto K and Edahiro T 1983b *Appl. Opt.* **22** 3850–8
Ioannidis ZK, Kadiwar R and Giles I P 1996 *J. Light wave Technol.* **14** 377–84
Johnson M 1979 *Appl. Opt.* **18** 1288–9
Johnstone W, Stewart G, Hart T and Culshaw B 1990 *IEEE J. Light wave Technol.* **8** 538–44
Kawachi M 1983 *Electron. Lett.* **19** 781–2
Kawachi M, Kawasaki BS, Hill KO and Edahiro T 1982 *Electron. Lett.* **18** 962–4
Kortenski T, Eftimov T and Vulkova T 1990 *Intern. J. of Optoelectronics* **5** 303–18
Krath KJ and Scholl B 1991 *IEEE Photonics Technol. Lett.* **3** 747–8
Kutsaenko V, Johnstone W, Lavretskii E and Rice J 1994 *IEEE Photonics Technology Lett.* **6** 1344–6
Laming R I and Payne D N 1989 *IEEE J. Light wave Technol.* **7** 2084–94
Lefevre HC 1980 *Electron. Lett.* **16** 778–80
Liu K, Sorin WV and Shaw HJ 1986 *Opt. Lett.* **11** 180–2
Luke DG, Mc Bride R, Burnett JG, Greenaway AH and Jones JDC 1995 *Optics Commun.* **121** 115–20.

Mermelstein MD 1986 *IEEE J. Light wave Technol.* **4** 449–53
Morishita K and Takashina K 1991 *IEEE J. of Light wave* **9** 1503–7
Muto S, Seki N, Ichikawa S and Ito H 1991 *Optics Commun.* **5** 273–5
Namihira Y and Wakabayashi H 1991 *J. Opt. Comm.* **12** 2–9
Nayar BK and Smith DR, 1983 *Opt. Lett.* **8** 543–5
Noda J, Okamoto K and Yokohama I 1987 *Fibre and Integrated Optics* **6** 309–30
Okoshi T, Cheng YH and Kikuchi K 1985 *Electron. Lett.* **21** 787–8
Pannell CN, Tatam RP, Jones JDC and Jackson DA 1988 *Fiber and Integrated Optics* **7** 299–315
Pickett CW, Burns WK and Villarruel CA 1988 *Opt. Lett.* **13** 835–7
Pilevar S, Thyagarajan K and Kumar A 1991a *J. Opt. Comm.* **12** 22–5
Pilevar S, Kumar A and Thyagarajan K 1991b *Opt. Comm.* **83** 31–6
Pleibel W, Stolen RH and Rashleigh SC 1983 *Electron. Lett.* **19** 825–6
Rashleigh SC and Marrone MJ 1983 *Electron. Lett.* **19** 850–1
Reichel V, Höfer B, Vobian J and Dultz W 1993 *Intern. J. of Optoelectronics* **8** 587–94
Risk WP and Kino GS 1986 *Opt. Lett.* **11** 48–50
Risk WP, Kino GS and Khuri-Yakub BT 1986 *Opt. Lett.* **11** 578–80
Rumbaugh SH, Jones MD and Casperson LW 1990 *IEEE J. Light wave Technol.* **8** 459–65
Rogers AJ 1985 *IEE Proc. PtJ* **132** 303–8
Sato T, Sun J, Kasahara R and Kawakami S 1999 *Opt. Lett.* **24** 1337–9
Stolen R H, Ashkin A, Pleibel W and Dziedzic J M 1985 *Opt. Lett.* **10** 574–5
Takada K, Okamoto K and Noda J 1986 *IEEE J Light wave Technol.* **4** 213–19
Tatam RP, Pannell CN, Jones JDC and Jackson DA 1987 *IEEE J. Light wave Technol.* **5** 980–5
Tatam RP, Hill DC, Jones JDC and Jackson DA 1988 *IEEE J. of Light wave Technol.* **6** 1171–6
Ulrich R 1979 *Appl. Phys. Lett.* **35** 840–2
Ulrich R and Simon A 1979 *Appl. Opt.* **18** 2241–51
Van Deventer O 1994 *IEEE J. of Light wave Technol.* **12** 2147–52
Villarruel CA, Abebe M and Burns WK 1983 *Electron. Lett.* **19** 17–18
Walker NG and Walker GR 1990 *IEEE J. Light wave Technol.* **8** 438–58
Yataki MS, Payne DN and Varnham MP 1985 *Electron. Lett.* **21** 249–51
Yokohama I, Noda J and Okamoto K 1987 *IEEE J. Light wave Technol.* **5** 910–15
Youngquist RC, Brooks JL, Risk WP, Kino GS, Shaw HJ 1985 *IEE Proc. PtJ* **132** 277–86
Yu T and Wu Y 1988 *Opt. Lett.* **13** 832–4

Zervas MN 1990 *IEEE Photonics Technol. Lett.* **2** 253–6
Zervas MN and Giles IP 1989 *Electron. Lett.* **25** 321–3

EXERCISES

4.1 – A mass of 100 kg is applied on 9 cm fibre length which has an outer diameter $2b = 250$ μm and a refractive index $n_1 = 1.46$.
    a – Evaluate the birefringence Δn_a ($K = 5 \times 10^{-12}$ SI, $S = 1.58$, $g = 10$ m / s).
    b – Calculate the phase difference for a wavelength $\lambda = 1.3$ μm.
4.2 – What is the intrinsic birefringence Δn_i in a fibre if the beat length is $L_b = 500$ μm at $\lambda = 1.3$ μm? Approximate the corresponding rotation angle θ if the fibre has the same strain as in exercise 4.1.
4.3 – Give a demonstration of equation (4.2).
4.4   **a** – In order to obtain with a fibre coil a phase shift of $\pi / 2$ at $\lambda = 1.3$ μm, what is the m value if the outer diameter of the fibre is 120 μm?
    b – If the number of turns is $N = 4$, calculate the radius R.

Chapter 5

DEVICES USING NON-LINEARITIES IN FIBRES

The classical treatment of propagation, light superposition, reflection, refraction, etc. assumes that there is a linear relationship between the electromagnetic field and the responding atomic system comprizing the medium. We know that an oscillatory mechanical device such as a weighted spring can be driven into a non-linear response with the application of large enough forces. In the same manner an intense light beam, such as a laser beam, can generate appreciable non-linear optical effects. The realm of non-linear optics encompasses these effects, in which high-intensity electric and magnetic fields play a dominant role.

A wide range of non-linear effects in optical fibres has been studied, and has found applications-in optical frequency converters, pulse compressors, switches, tunable source modulators, logic gates, optical fibre transistors, soliton lasers and light amplification. In optical telecommunications high transmitter power and low fibre losses are required. It has been recognized that non-linear optical processes probably present the practical limitation on the range and data transmission capacity of communications systems.

- Stimulated Brillouin scattering (SBS) is an interaction between light and sound waves in a fibre. It is responsible for frequency conversion and reverse propagation of light.
- Stimulated Raman scattering (SRS) is an interaction between light and vibrations of silica molecules. It also causes frequency conversion, and is responsible for excess attenuation of short wavelength channels in wavelength-multiplexed systems.
- Four-photon mixing (FPM) is analogous to third order intermodulation distortion, whereby two or more optical waves of different wavelengths mix to produce new optical waves at other wavelengths.

- Non-linearity of refractive index, known as the optical Kerr effect, has its origin in the third-order susceptibility $\chi^{(3)}$.
- The static Kerr effect, a quadratic variation of refractive index with applied voltage (and thereby electric field), is typical of several non-linear effects that have long been well known.
- Cross-phase modulation is an interaction, via the non-linear refractive index, between the intensity of one light wave and the optical phase of other light waves.

The electric fields associated with light beams from traditional sources are far too small for such behaviour to be easily observable. However, strong fields are readily obtainable with current laser technology. A good lens can focus a laser beam down to a spot having a radius of 20 μm or so, which corresponds to an area of roughly 1.256×10^{-9} m². A 200 megawatt pulse from a Q-switched ruby laser then produces a flux density of 1.59×10^{17} Wm^{-2}. The corresponding electric field amplitude E_0 is given by the intensity (or flux density) by the equation:

$$I = \frac{1}{2} v\varepsilon \langle E_0^2 \rangle = \frac{1}{2} n_1 \left(\frac{\varepsilon_0}{\mu_0}\right)^{1/2} \langle E_0^2 \rangle \tag{5.1}$$

so that

$$E_0 = 27.46 \left(\frac{I}{n_1}\right)^{1/2} \, Vm^{-1}$$

where n_1 is the refractive index, v the speed of propagation and ε the electrical permittivity in the medium. For $n_1 = 1.45$, the field amplitude is about 9.098×10^9 Vm^{-1}. This is more than enough to cause the ionization of air (roughly 3×10^6 Vm^{-1}) and about the same as the cohesive field of the electron in a hydrogen atom (5×10^{11} Vm^{-1}). The availability of fields greater than 10^{12} Vm^{-1} has made possible a wide range of important new non-linear phenomena and devices using these.

Stimulated Raman scattering and Stimulated Brillouin scattering limit the laser power which can be transmitted. Self phase modulation broadens the transmitted pulses, thus limiting the maximum data rate. However, the possibilities of using SBS and SRS (specially SRS) as optical amplifiers will be of increasing significance in the future. Today, optical fibre amplifiers and lasers are already used in telecommunications systems.

In order to achieve efficient non-linear optical effects in fibres there are three possibilities: the increase of optical intensity, the control of group velocity dispersion and the use of efficient non-linear optical material for fibres. High numerical aperture single-mode fibres with high optical intensity, single-mode fibres with low dispersion at the operating wavelength and $LiNbO_3$ (lithium niobate) single crystal fibres have been investigated (Sudo and Itoh 1990). The first possibility can find applications in optical amplifiers and lasers, the second in high-speed transmission or optical processing, and the third in devices such as second harmonic generators, modulators and optical parametric oscillators.

Several papers review both the detrimental and beneficial effects of optical non-linearities in fibres for telecommunications (Cotter 1987, Chraplyvy 1990, Shibata *et al.* 1990, Maeda *et al.* 1990, Lin 1986, Crosignani 1992).

Theoretical approaches to non-linearities in single-mode fibres, and theoretical and experimental studies of combined self phase modulation and stimulated Raman scattering have been made by several authors (Pocholle *et al.* 1990, Kean *et al.* 1987 and Kuckartz *et al.* 1987).

5.1 DEVICES BASED ON STIMULATED RAMAN SCATTERING (SRS)

5.1.1 Basic principle

When a photon of frequency v_i collides with and is absorbed by an atom, energy is transmitted to a bound electron, resulting in the excitation of the atom. If the frequency of the incident photon matches the excitation energy of the atom the absorption probability is greatest. In solids, liquids and dense gases absorption occurs over a band of frequencies, and the energy is generally dissipated by way of intermolecular collisions. In contrast, the excited atoms of a low-pressure gas can re-radiate a photon of the same frequency (v_i), in a random direction (resonance radiation). Accordingly, scattering is predominently at frequencies coinciding with the excitation energies of the atoms (see Figure 5.1).

Scattering can also occur, but with less likelihood at frequencies other than those corresponding to the stable energy levels of the atoms. A photon will be radiated without any appreciable time delay and most often with the same energy as the absorbed quantum (elastic or coherent scattering). There is also a phase relationship between the incident and scattered fields (Rayleigh scattering).

In some cases an excited atom does not return to its initial state after the emission of a photon. If the atom drops down only to an intermediate state, it emits a photon of lower energy than the incident primary photon; this is referred to as a *Stokes transition*. If the process

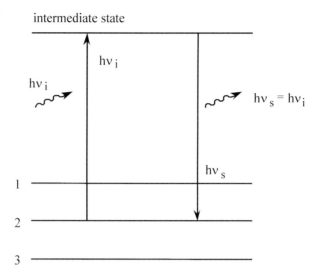

Figure 5.1. Radiation of a photon of the same frequency as the incident photon.

takes place rapidly (roughly 10^{-7} s) it is called *fluorescence*. If there is an appreciable delay (seconds, minutes or hours) it is known as *phosphorescence*.

If quasi-monochromatic light is scattered from a substance, it mainly consists of light of the same frequency. Yet it is possible to observe very weak additional components. They have higher and lower frequencies and the differences between these *sidebands* are characteristic of the material. This is the spontaneous Raman effect.

A molecule can absorb radiant energy in the far infrared and microwave regions, converting it into rotational kinetic energy. It can also absorb infrared photons, transforming the energy into a vibrational motion of the molecule. A quantum of vibrational energy is known as a *phonon*. It has a magnitude $h\nu$, where h is Planck's constant and ν is the frequency of the vibration. The molecule can also absorb energy in the visible and ultraviolet regions through the mechanism of electron transitions. Consider a molecule in some vibrational state (2 in Figure 5.2a). An incident photon of energy $h\nu_i$ is absorbed, raising the system to an intermediate state whereupon it immediately makes a *Stokes transition* $h\nu_s$, emitting a scattered photon of energy $h\nu_s$. The difference $h\nu_i - h\nu_s = h\nu_{32}$ goes into exciting the molecule to a higher vibrational energy level 3. Electronic and rotational excitation can also result. Furthermore, if the initial state is an excited one, the molecule after absorbing the photon $h\nu_i$ and

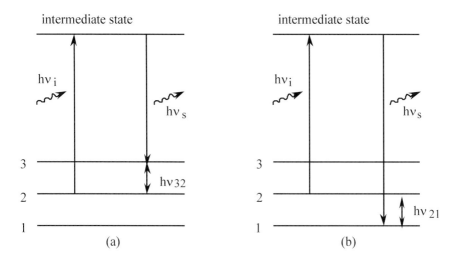

Figure 5.2. Scheme of spontaneous Raman effect.

emitting the photon $h\nu_s$ can drop back to an even lower state (1) making an *anti-Stokes transition* (see Figure 5.2b). In this case $h\nu_s > h\nu_i$. A vibrational energy $h\nu_{21}(=h\nu_s - h\nu_i)$ is converted into radiant energy. In either case the differences between ν_s and ν_i correspond to specific energy level differences for a substance.

Stimulated Raman scattering is a three-wave process in which the pump wave creates a frequency downshifted Stokes wave and a highly damped material excitation wave, which corresponds to the vibrational mode of SiO_2. At high enough optical intensities, the Raman scattering process becomes stimulated, so that the scattered light, called Stokes-shifted light, has laser-like characteristics of stimulated emission-directionality, high brightness and high coherence. The optical gain associated with the process namely amplification of the spontaneously scattered Stokes light, is called Raman gain, and is related to the spontaneous Raman spectrum and the scattering cross-section of the medium. For the oxides used in optical fibre fabrication such as silica, germanium and phosphorus glasses, the Raman spectra show frequency broadness (molecular vibration bands) rather than discrete lines, due to the amorphous nature of glass.

When an optical wave is propagating through a medium the photons are scattered. They produce a phonon by exciting molecular vibrations and a Stokes photon with a lower frequency (longer wavelength), so that the total energy is conserved.

Stimulated Raman scattering is depicted schematically in figure 5.3. Here, two photon beams are simultaneously incident on a

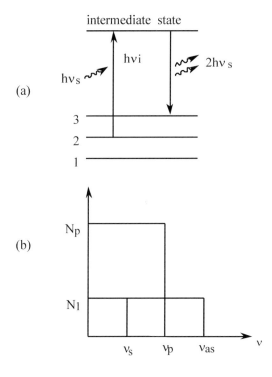
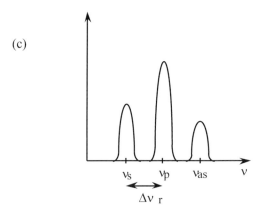

Figure 5.3. Scheme of stimulated Raman effect. $v_i = v_p$ pump frequency, v_s Stokes frequency, v_{as} anti-Stokes frequency. (a) Principle (energy diagram), (b) frequency division, (c) relative intensities.

molecule, one corresponding to a laser frequency v_i, the other having a scattered frequency v_s. The laser beam loses a photon hv_i, while the scattered beam gains a photon hv_s and is amplified. The remaining energy $hv_i - hv_s = hv_{12}$ is transmitted to the sample. The chain reaction in which a large part of the incident beam is converted into stimulated Raman light occurs above a particular threshold flux density of the exciting laser. It can occur in solids, liquids or dense gases (see Figure 5.3a).

At high temperatures some molecules are still vibrating, and the corresponding phonons may interact with the optical wave and produce an anti-Stokes wave with a higher frequency (shorter wavelength) (see Figure 5.3b). At ambient temperatures (near 300°K) the anti-Stokes intensity is much weaker than the Stokes intensity (see Figure 5.3c). In glasses the wave vector selection rules or phase matching conditions are lacking (in contrast to crystals) and the result is a broad Raman band, instead of a much narrower band corresponding to a well-defined frequency shift. Figure 5.4 shows the scattering emission of some oxide glasses used in optical fibre fabrication.

The applications of stimulated Raman scattering can be important: for example, CW Raman fibre lasers, tunable Raman oscillators, pulsed Raman lasers and fibre Raman amplifiers.

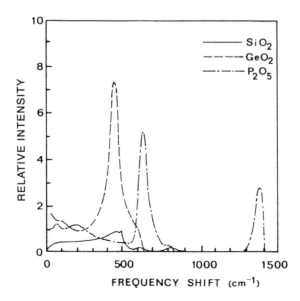

Figure 5.4. Spontaneous Raman scattering spectra of oxide glass used in fibre fabrication.

5.1.2 Amplification based on Raman effect

The propagation of a high-energy pumping laser beam through any optical fibre creates a Raman scattering spectrum, which extends to wavelengths longer than that of the pump. If power from a signal laser, with a wavelength that falls within the Raman spectrum created by the laser pump, is also injected into the fibre, then stimulated Raman scattering causes amplification of the signal. Large gains have been measured (Aoki et al. 1983, Nakamura et al. 1984). However, the crosstalk caused by SRS in single-mode fibre affects the performance (Tomita 1983).

Several papers have described calculations on amplified spontaneous scattering in fibres, concentrating on power limitations or on the dependence of total power on distance (Smith 1972, Stolen 1979, Pocholle et al. 1985, Aoki 1988, Auyeung and Yariv 1978, Dakss and Melman 1985, Aoki et al. 1986, Mochizuki et al. 1986, Nakashima et al. 1986) or on the gain saturation in fibre Raman amplifiers due to stimulated Brillouin scattering (Foley et al. 1989) or on review of theoretical and experimental studies (Aoki 1989, Stolen et al. 1984).

Consider a single-mode silica optical fibre with a length L (see Figure 5.5). The signal beam is launched at $z = 0$ (port 1 in coupler C_1). The pump light is injected at $z = 0$ for forward pumping (port 2) in coupler C_1 and at $z = L$ for backward pumping (port 4) in the coupler C_2. The amplified signal leaves from port 3. As the pump wave

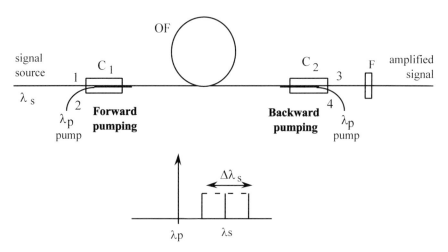

Figure 5.5. Basic configuration for Raman amplification. *OF* optical fibre, C_1 coupler used for forward pumping, C_2 coupler used for backward pumping, F filter to eliminate pump power in forward pumping.

propagates along the fibre, it is depleted by linear absorption and by conversion to Stokes photons.

For simplification, the signal beam is injected at a first Stokes wavelength, and the other non-linear processes are negligible. We can also neglect higher order Stokes waves which travel in an opposite direction to the signal because they do not reach the detector.

If $n_p(z)$ is the average number of pump photons, $n_{sj}(z)$ the jth order Stokes photon number, α_s the fibre transmission loss at the jth order Stokes wavelength, α_p the fibre transmission loss at pump wavelength and γ_j the Stokes gain coefficient for the jth mode, we have for forward stimulated Raman scattering:

$$\frac{dn_p(z)}{dz} = -\alpha_p n_p(z) - \sum_{j=1}^{q} \gamma_j n_p(z)(n_{sj}(z) + 1) \quad (5.2)$$

where q is the mode number. Each order $n_{sj}(z)$ obeys the equation:

$$\frac{dn_{sj}(z)}{dz} = -\alpha_s n_{sj}(z) + \gamma_j n_p(z)(n_{sj}(z) + 1) \quad (5.3)$$

The factor 1 in $(n_{sj}(z) + 1)$ corresponds to spontaneous emission. The term $n_{sj}(z)$ represents the amplified spontaneous Raman scattering (SRS). The analysis is carried out in the region where pump depletion is negligible. This region is important for signal amplification in that most of the gain is developed and no power loss is incurred due to a second Stokes generation. So the sum over j in (5.2) with respect to the $-\alpha_p n_p$ term is negligible. This uncouples the two equations. The solution to (5.2) is:

$$n_p(z) = n_0 \exp(-\alpha_p \cdot z) \quad (5.4)$$

where n_0 is the initial ($z = 0$) value of n_p. Assuming that $\alpha_s = \alpha_p = \alpha$ and substituting (5.4) for (5.3), a solution of equation (5.3) is (Dakss and Melman 1985, Aoki 1988):

$$n_{sj}(z) = n_{sj}(0) \exp(-\alpha z) \exp(\gamma_j n_0 z_e) + n_{sj}^*(z) \quad (5.5)$$

with $z_e = [1 - \exp(-\alpha z)]/\alpha$, $n_{sj}(0)$ is the initial injected average Stokes photon number (at $z = 0$), $\exp(\gamma_j n_0 z_e)$ is the Raman gain factor; the term $n_{sj}^*(z)$ represents the amplified spontaneous Raman scattering (ASRS).

The Stokes gain coefficient is supposed to be the same for all the modes in the interaction ($\gamma_j = \gamma_0$). As pump depletion due to Raman interaction is negligible, the higher order ($j \geqslant 2$) Stokes generation may be neglected. If we also suppose ASRS negligible, we have for the terms of (5.5):

$$\frac{dn_s}{dz} = -\alpha n_s + \gamma_0 n_0 \exp(-\alpha z) \times n_s \quad (5.6)$$

and

$$n_s = n_s(0) \exp[-\alpha z + \gamma_0 n_0 z_e] \quad (5.7)$$

The term $\exp[-\alpha z]$ represents the fibre transmission loss for the signal whereas the term $\exp(\gamma_0 n_0 z_e)$ represents the amplification factor in the fibre.

In order to take into account the polarization scrambling between the pump and signal waves, a polarization scrambling factor K is included in the exponential term. Complete polarization scrambling is expressed by setting $K = 2$ and the Raman gain is:

$$G = \exp(\gamma_0 n_0 z_e / K) \quad (5.8)$$

Equation (5.7) can be written in terms of Stokes power per unit frequency interval using:

$$P_s(\nu) = h\nu n_s$$

and also pump power P_p rather than in terms of average numbers of photons. The number of pump photons per unit length is n_p/z and the pump power is:

$$P_p = (h\nu_p V_g n_p)/z \quad \text{and} \quad P_p(0) = (h\nu_p V_g n_0)/z$$

with V_g the pump group velocity in the fibre.

Theoretical values of P_p and P_s in function of length z are given in Figure 5.6.

We can write the equations (5.2), (5.6) in terms of Stokes and pump power rather than average number of photons and the coupled differential equations governing the amplified signal along the single-mode fibre are given, in forward pumping, by:

$$\frac{dP_p(z)}{dz} = -\left(g_r \frac{\omega_p}{\omega_s} \frac{P_s(z)}{A_e} + \alpha_p\right) P_p(z)$$

$$\frac{dP_s(z)}{dz} = \left(g_r \frac{P_p(z)}{A_e} - \alpha_s\right) P_s(z) \quad (5.9)$$

P_p and P_s are the pump and signal powers, A_e is the effective interaction area, g_r is the Raman gain coefficient. $g_r = 9.2 \times 10^{-14}$ mW^{-1} for a pure SiO$_2$ core fibre at 1.0 μm pump wavelength.

The Raman gain spectrum of silica glass is shown in Figure 5.7.

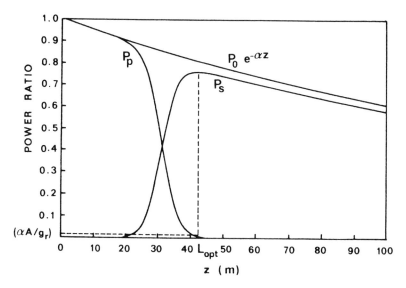

Figure 5.6. Theoretical values of P_p (z) and P_s (z) for $\alpha = 5 \times 10^{-5}$ cm [After Auyeung and Yariv 1978].

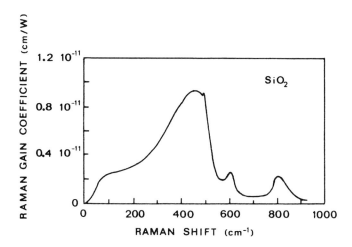

Figure 5.7. Raman gain coefficient of a typical silica glass fibre [After Lin 1986].

A_e can be expressed as an overlap integral accounting for the spot size mismatch between the pump and Stokes waves in the fibre. However, in single-mode fibres, the effective interaction area A_e is in the order of the core area (or effective mode spot size), which can be less than

10 μm in diameter. Then the effective interaction area is $A_e = \pi \omega_e^2$ where ω_e is the effective core radius.

In the linear and small amplification regime, assuming negligible pump depletion, the amplified signal power is obtained by the integration of equation (5.9) along the length L, using $P_p = P_p(0) \exp -\alpha_p \cdot L$ (Stolen 1979, Davey et al. 1989):

$$P_s(L) = P_s(0)[\exp(-\alpha_s L) \cdot G(\lambda)]$$

$$G(\lambda) = \exp\left[g_r P_p(0) \frac{L_e}{A_e K}\right]$$

$$G[dB] = 4.34 \cdot \frac{g_r P_p(0) \cdot L_e}{A_e K} \tag{5.10}$$

$$L_e = [1 - \exp(-\alpha_p L)]/\alpha_p$$

G is the Raman gain. $K = 1$ for a polarization-maintaining (PM) fibre, or $K = 2$ for a non-PM fibre. $P_s(0)$ represents the input signal at the Stokes wavelength. For a very long fibre $L_e \cong 1/\alpha_p$ whereas for short fibres $L_e = L$. This quantity is the effective interaction length and shows the effect of pump power attenuation along the fibre. For a long fibre, the pump power level for which the gain equals the loss (i.e. $P_s(L) = P_s(0)$) can be estimated from equation (5.10) as follows:

$$P_p(0) = \alpha_s \frac{A_e K}{g_r} \frac{L}{L_e} = \alpha_s \alpha_p \frac{A_e}{g_r} KL$$

For low-loss fibres the effective interaction length L_e could be tens of kilometres long.

The SRS threshold condition for significant SRS from noise (i.e. in the absence of input radiation in the Stokes frequency band) is $GL_e = 16$.

Figure 5.8 shows the theoretical Raman gain as a function of fibre length for several fibre losses.

When the input signal power is as high as the pump power, the hypothesis of negligible pump depletion and higher order Stokes generation is not fulfilled and Raman gain will become saturated.

In backward pumping the signs in the right part of equation (5.2) are modified and we obtain:

$$n_p(z) = n_p(L) \exp\left[-\alpha_p(L - z)\right]$$

By pumping a single-mode silica fibre, Raman amplification of laser diode light is obtained. However, CW Raman amplification is severely limited by the pump power depletion due to stimulated Brillouin scattering.

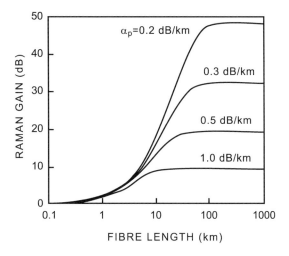

Figure 5.8. Theoretical Raman Gain as a function of fibre length and pump loss $g_r = 6.7 \times 10^{-14}$ m/W; input pump power 1.6 W; core diameter 10 μm [After Aoki 1988].

Fibre Raman amplifier properties and applications to long distance optical communications have been studied by Spirit *et al.* (1990), for example.

5.1.3 Raman gain in a re-entrant fibre loop

In a passive mode operation, the impulse response of a re-entrant fibre loop is a train of exponentially decaying pulses (see Figure 5.9a). These pulses are delayed by the loop transit time $\Delta\tau_l$. If I_i is the intensity of the input signal pulse with duration smaller than $\Delta\tau_l$, the fibre coupler splits this intensity with a coupler splitting ratio CR.

So a fraction $I_0 = CR I_i$ is tapped out of the loop and $(1-CR) I_i$ remains in the loop. If L is the interaction length (or actual fibre length) and α_s the fibre loss per unit length, the signal loses a fraction $\exp(-\alpha_s L)$ after propagation along the fibre. Because of the interaction of the process, a train of decaying pulses $I_1, I_2 \ldots I_4$ is obtained at the output port of the loop. Between two consecutive pulses I_{n+1} and I_n, we have:

$$\frac{I_1}{I_0} = \frac{(1-CR)^2}{CR} \exp(-\alpha_s L) \qquad \frac{I_{n+1}}{I_n} = CR \exp(-\alpha_s L) \qquad (5.11)$$

where α_s is the attenuation coefficient at the signal wavelength (Desurvire *et al.* 1986, Desurvire *et al.* 1985).

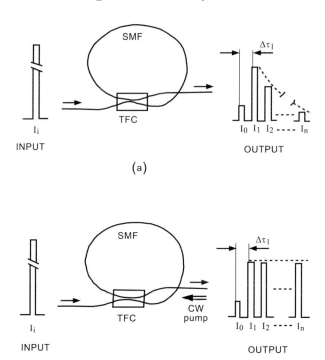

Figure 5.9. Scheme of a re-entrant fibre loop: (a) in the passive mode operation, (b) in the active mode operation. TFC tunable fibre coupler, SMF single mode fibre [After Desurvire et al. 1986].

In an active mode operation, a distributed Raman gain is provided in the fibre at the signal frequency v_s by coupling in the loop a pump power with frequency v_p (see Figure 5.9b).

For the CW pumping operation and in a linear amplification regime, there is no major difference between the forward and backward Raman gain. If CR_p is the coupler splitting ratio for pump, and P_{pi} the input pump power, the effective pump power P_p is given by the incoherent sum of the re-circulating pump power order:

$$P_p = P_{pi}(1 - CR_p) \sum_{n=0}^{\infty} (CR_p \exp(-\alpha_p L))^n = P_{pi} \frac{(1 - CR_p)}{1 - CR_p \exp(-\alpha_p L)} \quad (5.12)$$

α_p being the attenuation coefficient at v_p.

In the undepleted pump approximation, the steady state Raman gain G is given by equation (5.10) if $K = 1$.

The Raman gain compensates the signal losses when the following condition is satisfied (CR_s is the coupling ratio signal):

$$\frac{I_{n+1}}{I_n} = CR_s \exp(-\alpha_s L) = \frac{1}{G}$$

Using equation (5.10) we have:

$$G = \frac{1}{CR_s} \exp(\alpha_s L) = \exp\left(g_r \frac{L_e}{A_e} P_p(0)\right)$$

thus

$$-\ln CR_s + \alpha_s L = g_r \frac{L_e}{A_e} P_p(0) \tag{5.13}$$

From equations (5.12) and (5.13) with $P_{pi} = P_c$ we obtain the initial input pump power P_c corresponding to a critical Raman gain $G = G_c$ for a length L of fibre.

$$P_c = \frac{A_e}{g_r L_e} \frac{1 - CR_p \exp(-\alpha_p L)}{1 - CR_p} [\alpha_s L - \ln CR_s]$$

and

$$G_c = \frac{1}{CR_s} \exp(\alpha_s L) \tag{5.14}$$

So a single-mode fibre ring Raman laser could be fabricated by using a fibre coupler of high multiplexing effect (Desurvire et al. 1987).

5.2 DEVICES BASED ON STIMULATED BRILLOUIN SCATTERING (SBS)

5.2.1 Basic principle

Brillouin scattering is the interaction of light with acoustic waves in solids or liquids. The interaction occurs through the modulation of the refractive index of the medium in the alternating areas of compression and rarefaction of the acoustic wave. The acoustic wave forms a phase grating moving at the speed of sound in the medium. This grating can diffract an optical wave, changing its direction of propagation and its frequency via the Doppler effect.

The non-linear process in fibres that has the largest gain and hence the lowest light threshold is *stimulated Brillouin scattering* (SBS). It is a three-wave process involving the incident light-wave (pump), a scattered light wave (Stokes) and a generated acoustic wave (Cotter 1982, Ippen and Stolen 1972, Pocholle et al. 1990, Montes et al. 1989, Henry 1992). The pump beam is converted into Stokes light of longer wavelength along with phonons. The pump beam thus creates a pressure wave in the medium owing to electrostriction, and the resulting variation in density changes the optical susceptibility. In this process a significant proportion of the optical power travelling in the fibre may be converted into the Stokes wave which travels backwards, and is shifted to a lower frequency v_s with respect to the pump of frequency v_p by an amount equal to the acoustic frequency v_B. (In a single-mode optical fibre, the only possible change of direction is back-reflection). The frequency change v_B is given by:

$$v_B = \pm 2n_1(V_a/c)v_p = \pm 2n_1 \frac{V_a}{\lambda_p} = v_p - v_s \tag{5.15}$$

where n_1 is the core refractive index, V_a is the speed of sound in the medium, c is the speed of light in empty space and v_p is the frequency of the pump beam. The minus sign refers to scattering from an acoustic wave propagating in the same direction as the pump wave, and the plus sign corresponds to scattering from a wave counterpropagating. When the frequency shift is negative, energy is added to the acoustic wave; this is called *Stokes scattering*. When the shift is positive, energy is removed from the acoustic wave; this is called *anti-Stokes scattering*. In fused silica, taking the values, $n_1 = 1.44$ and $V_a = 5960$ ms^{-1} (see Figure 5.10a). We have approximately:

$$v_B \cong \frac{-17.3}{\lambda_p} \tag{5.16}$$

with v_B in GHz and λ in μm. For example, for $\lambda = 1.55$ μm we have $v_B = 11.1$ GHz.

Brillouin linewidth Δv_B, depending on fibre parameters, can be given approximately by:

$$\Delta v_B \cong 0.1412[v_B]^2 \ (MHz) \tag{5.17}$$

with v_B in GHz. The optical bandwidth Δv_B for SBS in silica is about 20 MHz at 1.55 μm and varies as λ^{-2}.

Stimulated Brillouin scattering could be a limiting factor in the data transmission capacity of communication systems (Montes et al. 1989, Hirose et al. 1991).

5.2.2 Fibre Brillouin amplifiers

The principle of operation of scattering amplifiers is the amplification of an external signal by coupling into the fibre a certain amount of the

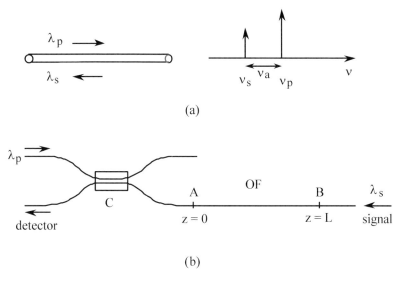

Figure 5.10. (a) Schematic of Brillouin scattering, (b) Schematic of a fibre Brillouin amplifier. S signal source, D detector, C coupler, OF optical fibre.

pump power. It provides a distributed gain at the signal wavelength. Stimulated Brillouin amplification is an intrinsic process in which the pump beam can amplify counterpropagating signal light of slightly lower frequency. Several applications have been proposed, and several papers describe the properties of fibre Brillouin amplifiers (Bayvel et al. 1989, Botineau et al. 1994, Tkach and Chraplyvy 1989). Stimulated Brillouin amplification has been used for channel selection in a dense-packed WDM direct detection light-wave experiment (Chraplyvy and Tkach 1986, Lee et al. 1992a,b). A coherent three wave SBS (stimulated Brillouin scattering) model and an adiabatic model can be used (Botineau et al. 1989, Botineau et al. 1988). The forward propagating pump beam of amplitude E_p at frequency ω_p, coupled with the thermal phonon fluctuation F of the core medium, stimulates a counterpropagating Stokes wave E_s at a frequency ω_s. This wave E_s is downshifted by the frequency $\omega_B = 2V_a\omega_p n_1/c$ where n_1 is the fibre refractive index. We can write:

$$E_p(z,t) = E_{0p}(z,t) \exp\left[i(k_p z - \omega_p t)\right]$$

$$E_s(z,t) = E_{0s}(z,t) \exp\left[i(\omega_s t - k_s z)\right]$$

$$S(z,t) = S_0(z,t) \exp\left[i(k_j z - \omega_B t)\right] \quad (5.18)$$

with

$$\omega_s = \omega_p - \omega_B \quad \text{and} \quad \mathbf{k}_j = \mathbf{k}_p - \mathbf{k}_s$$

$S(z,t)$ is induced by electrostriction. In optical fibre experiments and in small coupling cases (the flux $\phi \leq \phi_c = 6.10^{12}$ W/cm^2 in silica) several approximations can be made. With a slowly varying envelope approximation for the wave, the complex amplitudes E_{op}, E_{os}, S_o vary slowly in time and space (the frequencies ω_i are high compared with the time variation of the amplitudes).

Neglecting the variation with time, the evolution equations for the powers $P_i = \frac{1}{2} n_1 \varepsilon_0 c |E_{0i}|^2 * A_e$ ($i = p$ or s) are:

$$\frac{dP_p(z)}{dz} = -\alpha P_p(z) - \frac{g_B P_s(z)}{A_e} P_p(z)$$

$$\frac{dP_s(z)}{dz} = +\alpha P_s(z) - \frac{g_B P_p(z)}{A_e} P_s(z) \tag{5.19}$$

where A_e is effective mode area and α corresponds to the linear optical attenuation of the pump. g_B is the Brillouin gain coefficient:

$$g_B = \frac{2\pi n_1^7 p_{12}^2 K}{c \lambda_p^2 \rho V_a \Delta v_B} \tag{5.20}$$

where $p_{12} (= 0.283)$ is the longitudinal elasto-optic coefficient, Δv_B the linewidth for spontaneous Brillouin scattering, and K the polarization factor which is unity for a fibre which maintains polarization and equal to one-half otherwise. ρ is the material density ($\rho = 2.210 \times 10^3$ Kg m^{-3}). It is assumed that the laser line width is small compared with Δv_B. The value of g_B is approximately 4.10^{-11} mW^{-1}.

Figure 5.10b shows a schematic diagram of a fibre Brillouin amplifier (FBA), configured as a preamplifier. A coupler is used to inject the pump and to allow the signal to reach the receiver. Signal power $P_s(L)$ is injected at the near end, and pump power $P_p(0)$ is injected into the fibre, after the coupler, at the far end. It is assumed that the pump is attenuated only by the fibre loss, which means that the amplifier is unsaturated:

$$P_p(z) = P_p(0) \exp(-\alpha z)$$

At the end of a fibre length L, the signal power P_s at the wavelength λ_s of the scattered lightwave can be amplified from the pump power P_p. From equation (5.19) we have, after integration between $z = 0$ and $z = L$:

$$P_s(0) = P_s(L) \exp(-\alpha L) \exp\left(g_B \frac{P_p(0)}{A_e} L_e\right)$$

Devices based on stimulated Brillouin scattering (SBS)

This equation shows amplification, which can be quite large when

$$\frac{g_B \cdot P_p(0)}{(A_e)} \cdot L_e > \alpha \cdot L$$

is the effective cross section area determined from the mode spot size ω_0. L_e, the effective interaction length, taking into account the linear pump absorption, is given in equation 5.10:

$$L_e = [1 - \exp(-\alpha_p L)]/\alpha_p$$

Sometimes, the equations are written with G, which is the SBS gain factor and is defined by:

$$G = g_B \frac{P_p(0)}{A_e}$$

The maximum laser power P_c (critical power) that can be launched into the fibre before SBS becomes detectable (i.e., the level at which it begins to degrade communication systems performance) is determined by:

$$P_c \approx 21 \frac{A_e K}{g_B L_e}$$

i.e., $G_c L_e = 21$ if the polarization factor $K = 1$.

Maximum SBS gain will occur for pump lasers with linewidths Δv_p less than 20 MHz. If the linewidth of the laser Δv_P is much larger than 20 MHz, SBS gain decreases in proportion to the ratio $\Delta v_B/\Delta v_P$, and the gain g_B becomes $g_B = g_{BM} \Delta v_B/\Delta v_P$ where g_{BM} is the maximum steady state Brillouin gain.

For the strong gain peak, the SBS threshold power P_c for CW operation is given by the expression:

$$P_c(CW) = 21 \left(\frac{A_e K}{g_B L_e}\right) \left(\frac{\Delta v_B + \Delta v_p}{\Delta v_B}\right)$$

For a typical fibre loss value at 1.5 μm, say $\alpha = 0.2$ dB km^{-1}, $L_e = 21.7$ km in a long fibre.

Typically for an 8 μm diameter fibre core, the gain at 1.5 μm is about 36 dB mW^{-1} over a bandwidth of 17 MHz. For system applications it is necessary to broaden the gain bandwidth (Olsson and Van der Ziel 1986).

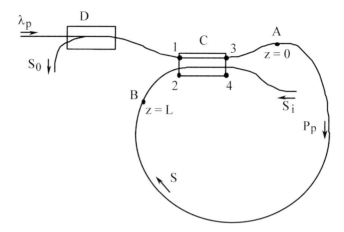

Figure 5.11. Scheme of a Brillouin laser/amplifier: C is an X coupler, D a wavelength filter.

5.2.3 Brillouin laser based on a fibre ring resonator

A stable train of compressed Stokes pulses can be obtained in a stimulated Brillouin fibre ring laser by periodically interrupting a pump beam with an intra-ring cavity acousto-optic modulator (Botineau et al. 1989). The possibility of using an optical fibre Brillouin ring laser for inertial sensing has been tried (Thomas et al. 1980).

A Brillouin laser/amplifier based on the all-fibre ring resonator is made with a loop of length L and a tunable directional coupler (see Figure 5.11). Initially we consider the operation of the Brillouin laser in the absence of an external Stokes signal S_i. The pump field E_p circulates in the resonator to amplify the counter-propagating spontaneous Stokes noise. The pump light propagates in one direction, from $z = 0$ to $z = L$ in the loop and the Stokes light travels in the opposite direction (from $z = L$ to $z = 0$) with a period $\tau = n_1 L/c$.

The coupled equations, for the slowly varying complex amplitudes of the pump and Stokes waves, are:

$$\begin{aligned}\frac{\partial E_p^{m+1}}{\partial z} &= -\frac{1}{2}\left(\gamma' + g_B |E_s^m|^2\right) E_p^{m+1} \\ \frac{\partial E_s^{m+1}}{\partial z} &= \frac{1}{2}\left(\gamma' - g_B |E_p^{m+1}|^2\right) E_s^{m+1}\end{aligned} \quad (5.21)$$

where E_p and E_s are the pump and Stokes field amplitudes and g_B is the peak Brillouin gain coefficient in m W^{-1}. The index m (where $m = 1, 2, 3 \ldots$) refers to the number of transit periods, around the loop. The propagation of the field once around the loop is considered as one circulation. Here $\gamma' = (\alpha + j\beta)$, where α is the intensity attenuation coefficient and $\beta/2$ the propagation coefficient of the pump and Stokes fields. Since the Brillouin shift is very small (approximately 20 GHz at 830 nm), we consider the magnitude of the total phase over the length L, $\beta L/2$ to be the same for the pump and Stokes waves.

The time taken for the light to propagate along the length of the directional coupler is insignificant compared with the time along the length of the loop. The build-up of the pump and Stokes fields, as a function of the number of circulations, effectively gives the transient response of the device.

Because the pump and the Stokes waves propagate in opposite directions, the boundary conditions for them have to be applied on the opposite boundaries. This means that equations become impossible to solve analytically, and generally numerical calculations are carried out using the assumption that pump depletion is negligible. However, pump depletion is an important factor in the calculations of the lasing threshold, of the conversion efficiency of the Brillouin laser, in the evaluation of the SNR (signal-to-noise ratio) and of gain of Brillouin amplifiers. In addition, the fact that we are considering the coherent addition of fields inside the resonator further complicates the solutions, necessitating numerical analysis of considerable complexity (Bayvel et al. 1989, Bayvel and Giles 1990, Stokes et al. 1982).

In order to simplify the solutions the following approximation is made. Since the resonator loop length L is in general short (<15 m), the total Brillouin gain over one loop transit is small. So:

$$E_p^{m+1}(L) = E_p^{m+1}(0) \exp\left\{-\frac{1}{2}\left(\gamma' + g_B |E_s^m(0)|^2 \cdot L\right)\right\}$$

The factor $g_B|E_s^m(0)|^2$ in the exponential term now accounts for the pump depletion. Similarly, for the Stokes wave, integration of equation (5.21) yields:

$$E_s^{m+1}(L) = E_s^{m+1}(0) \exp\left\{+\frac{1}{2}\left(\gamma' - g_B |E_p^{m+1}(0)|^2 \cdot L\right)\right\}$$

The coupler radiation loss is included as a lumped loss in the expression for the coupling coefficient. It can be seen that the boundary conditions are, for the pump wave:

$$E_p^{m+1}(0) = jCE_p^m(L) + E_0$$

where $E_0 = (1 - C^2)^{1/2} E_{in}$, and C is the field coupling coefficient, and, from the above, C includes the coupler radiation loss.

In the case of polarization-maintaining fibre, a fibre Brillouin ring laser has been constructed and operated without instability due to

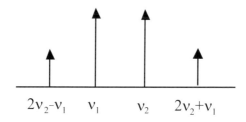

Figure 5.12. Scheme of four wave mixing with two injected waves at frequencies v_1 and v_2.

interaction between the polarization lateral modes. In this case the fibre was set with 90° polarization axis rotation at the splice (Tanaka and Hotate 1995).

Superluminous quasi-solitons may be observed in a Brillouin fibre ring laser if the acoustic wave has inertia with a long spontaneous decay time. In this case, description of the different wave interactions must take into account the dynamic behaviour of the stimulated Brillouin backscattering (Picholle et al. 1991).

5.3 PARAMETRIC FOUR WAVE MIXING

In this technique, several optical frequencies are launched into a fibre simultaneously (see Figure 5.12) (Hill et al. 1978, Stolen et al. 1981, Stolen and Bjorkholm 1982, Kikuchi and Lorattanasane 1994, Inoue and Toba 1992, Reis et al. 1995).

Three waves of frequencies ω_1, ω_2 and ω_3 can interact via the third-order non-linear susceptibility to generate a fourth wave of frequency ω_4, which follows the energy conservation relation:

$$\omega_4 = \omega_1 + \omega_2 - \omega_3 \tag{5.22}$$

A significant transfer to the new frequency ω_4 is possible only if momentum conservation (phase matching condition) is satisfied. Conservation of momentum requires:

$$\Delta \mathbf{k} = \mathbf{k}_1 + \mathbf{k}_2 - \mathbf{k}_3 - \mathbf{k}_4$$

where $\mathbf{k}_i$ are propagation vectors in the fibre.

Parametric mixing can also occur between two input waves ω_p and ω_s, in which case the energy conservation relation becomes:

$$2\omega_p = \omega_s + \omega_{as}$$

where $\omega_p (= \omega_1 = \omega_2)$ is the pump frequency, ω_s is the signal frequency and ω_{as} is the generated frequency. The phase matching condition is:

$$\Delta \mathbf{k} = 2\mathbf{k}_p - \mathbf{k}_s - \mathbf{k}_{as}$$

In a single-mode fibre the propagation vector at a given frequency is the sum of two contributions. The first, $\Delta \mathbf{k_m}$, is determined by the bulk dispersion of the material. The second, $\Delta \mathbf{k_w}$, depends on the dispersive characteristic of the waveguide itself (guiding structure and geometrical parameters). The total phase mismatch is the sum of these two parts:

$$\Delta \mathbf{k} = \Delta \mathbf{k_m} + \Delta \mathbf{k_w}$$

$$\Delta k_m = (n_{as}\, \omega_{as} + n_s\, \omega_s - 2n_p\, \omega_p)/c$$

$$\Delta k_w \cong (B_{as}\, \omega_{as} + B_s\, \omega_s - 2B_p\, \omega_p)\Delta n/c$$

where

$$B_j = \frac{n_{ej}^2 - n_2^2}{n_1^2 - n_2^2} = \frac{(\beta_j/k_j)^2 - n_2^2}{n_1^2 - n_2^2}$$

is the normalized propagation constant, n_{ej} is the effective index for each wavelength j (due to guiding effect and defined with the help of equation 1.25 Chapter 1), subindex s corresponds to the Stokes signal, subindex as to the anti-Stokes signal, and p to the pump signal; c is the speed of light in empty space. Δn is the core/cladding index difference.

A useful approximation to the contribution Δk_m is:

$$\Delta k_m = 2\pi \lambda D(\lambda) \Omega^2 \qquad (5.23)$$

with $\Omega = \dfrac{|\omega_p - \omega_s|}{2\pi c}$ (cm^{-1}) and $D(\lambda) = \lambda^2 \dfrac{\partial^2 n_{ej}}{\partial \lambda^2}$

$D(\lambda)$ is the group velocity dispersion and must not be too small for the validity of equation (5.23).

Tunable wavelength conversion is possible using two pump beams. A signal beam is converted from one arbitrary frequency to another compatible with the zero dispersion wavelength of the fibre. Tuning one of the pump beam frequencies permits selective conversion (Inoue 1994).

In single-mode non-polarization-maintaining fibres, phase matching can be obtained when the pump wavelength λ_p is slightly greater than the wavelength λ_0 for zero group velocity dispersion. The small negative Δk_m is compensated by Δk_w. A second case is where the frequency shift ($\omega_p - \omega_s$) is sufficiently small to obtain a coherence length L_{co} for parametric interaction ($L_{co} = 2\pi/\Delta k$) similar to or greater than the fibre length. In polarization-maintaining fibre, phase matching is possible by using a large birefringence (Stolen et al. 1981). Stimulated four-photon mixing in birefringent fibres with the pump wave propagating in two orthogonally polarized modes has been investigated by Ovsyannikov et al. (1991).

Four-wave mixing can be used in multi-amplifier and high-speed WDM systems (Inoue and Toba 1995, Tkach et al. 1995) or for wavelength conversion using a loop configuration (Yu and Jeppesen 2000).

5.4 KERR NON-LINEARITIES IN OPTICAL FIBRES – SOLITONS

5.4.1 Basic principle

Lasers producing high-intensity short-duration pulses are now readily available. It is possible with such lasers to investigate potentially useful non-linear effects in fibre. The electromagnetic field of a light wave propagating through a medium exerts forces on the loosely bound outer or valence electrons. Ordinarily, these forces are quite small and in a linear isotropic medium the resulting electric polarization **P** is parallel to and directly proportional to the applied field **E**. Indeed, the polarization follows the field; if the latter is sinusoidal, the former will be sinusoidal as well. Consequently, one can note:

$$\mathbf{P} = \varepsilon_0 \chi \mathbf{E} \qquad (5.24)$$

where χ is a dimensionless constant known as the electric susceptibility and ε_0 the permittivity. A plot of **P** versus **E** is a straight line. In the extreme case of very high fields we can expect that **P** will become saturated, i.e., it cannot increase linearly indefinitely with **E** (just as in the case of ferromagnetic materials where the magnetic moment becomes saturated at fairly low values of **H**). Thus we can anticipate a gradual increase in the usually insignificant non-linearity as **E** increases. Since in the simplest case of an isotropic medium the directions of **P** and **E** coincide, we can express the polarization more effectively as a polynomial expansion:

$$\mathbf{P} = \varepsilon_0 \left(\chi \mathbf{E} + \chi_2 \mathbf{E}^2 + \chi_3 \mathbf{E}^3 + \cdots \right) \qquad (5.25)$$

The usual linear susceptibility χ is much greater than the coefficients of the non-linear terms χ_2, χ_3, etc. and hence the latter contributes noticeably only at high-amplitude fields. If the form of the light wave incident on the medium is:

$$\mathbf{E}(t) = \mathbf{E_0} \sin \omega t$$

The resulting electric polarization is:

$$\mathbf{P} = \varepsilon_0 \chi \mathbf{E_0} \sin \omega t + \varepsilon_0 \chi_2 \mathbf{E_0}^2 \sin^2 \omega t + \varepsilon_0 \chi_3 \mathbf{E_0}^3 \sin^3 \omega t + \cdots$$

This can be rewritten as:

$$\mathbf{P} = \varepsilon_0 \chi \mathbf{E_0} \sin \omega t + \frac{\varepsilon_0 \chi_2}{2} \mathbf{E_0}^2 (1 - \cos 2\omega t)$$
$$+ \frac{\varepsilon_0 \chi_3}{4} \mathbf{E_0}^3 (3 \sin \omega t - \sin 3\omega t) + \cdots$$

As the sinusoidal wavefront sweeps through the medium, it creates what might be thought of as a polarization wave, i.e., an undulating redistribution of charge within the material in response to the field. If only the linear term is effective, the electric polarization wave would correspond to an oscillatory current following along with the incident light. The light thereafter reradiated in such a process would be the usual refracted wave propagating with a reduced speed v and having the same frequency as the incident light. In contrast, the presence of higher order terms implies that the polarization wave has the same harmonic profile as the incident field representation of the distorted profile of $\mathbf{P}$.

The non-linear relationship between the electrical polarization $\mathbf{P}$ and the electrical field strength $\mathbf{E}$ in a dielectric optical fibre induces non-linearity of the refractive index. The third-order susceptibility χ_3 is responsible for the optical Kerr effect. The second-order susceptibility χ_2 vanishes in fibres due to the inversion symmetry of fused silica material. The physical process underlying the appearance of optical solitons is the Kerr effect, which leads to the self-phase-modulation (SPM) of high-power light pulses propagating over a long silica fibre. It can be characterized by an intensity dependence of the refractive index (Stolen and Lin 1978). If I (given by equation 5.1) is the beam intensity and n_0^* the non-linear refractive index:

$$n(I) = n_1 + n_0^* I = n_1 + \frac{n^*}{2} |E(t)|^2 \qquad (5.26)$$

$$\text{with } n^* = \frac{n_1 n_0^*}{\mu_0 c}$$

where n_1 is the refractive index at low intensity, n^* is expressed in (m^2 V^{-2}), n_0^* is expressed in (m^{2}W^{-1}) and c in (ms^{-1}). For pure silica n^* is 1.15×10^{-22} m^{2}V^{-2} and $n_0^* = 3.2 \times 10^{-20}$ m^2/W^{-1}.

A pulse with the intensity envelope $E^2(t)$ will induce a non-linear refractive index variation: $\Delta n(t) = n^* E^2(t)/2$. Consequently, the SPM $\Delta\Phi(t)$ of the wave packet propagating along a fibre length L is expressed by:

$$\Delta\Phi(t) = (2\pi L/\lambda) \, \Delta n(t) = (\pi L \, n^*/\lambda) \, E^2(t)$$

So the phase shift is proportional to the fibre length and to the intensity in a lossless fibre. The approximate frequency shift at time t is given by the time derivative of the phase perturbation, which is proportional to the power:

$$\delta\omega(t) = -\frac{\delta\Delta\Phi}{\partial t} = -\frac{2\pi L}{\lambda}\frac{\delta\Delta n}{\partial t}$$

If the initial pulse is of the form $E(t) = A \operatorname{sech}(t)$, the instantaneous frequency shift within the pulse is given by:

$$\delta\omega(t) = -\frac{\pi L}{\lambda} n^* A^2 \frac{\partial}{\partial t}\left[(\operatorname{sech}(t))^2\right]$$

This equation shows that a pulse develops a chirp proportional to the distance L (Blow and Doran 1987) (see Figure 5.13). This phenomenon can be used in pulse compression.

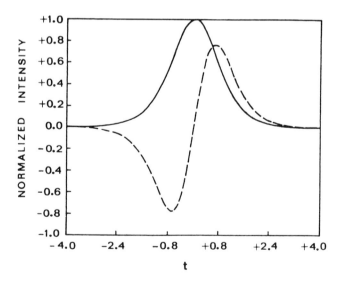

Figure 5.13. Pulse intensity (solid line), resultant chirp (broken line) [After Blow and Doran 1987].

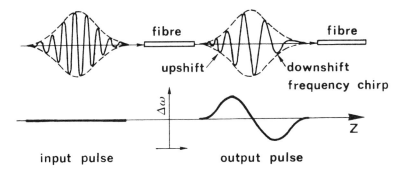

Figure 5.14. Scheme of self-phase modulation in fibre [After Veith 1988].

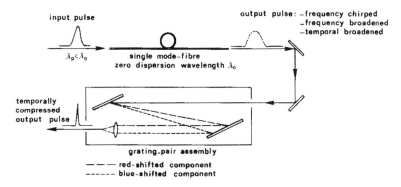

Figure 5.15. Optical pulse compression scheme using fibre and gratings [After Veith 1988].

A pulse with a Gaussian profile propagating along a silica fibre will undergo a frequency downshift at the leading edge and a frequency upshift at the trailing edge (see Figure 5.14) (Veith 1988).

Several papers review the aspect of SPM in optical fibre, applications (Veith 1988, Blow and Doran 1987, Cotter 1987) or examine the optical Kerr effect in long fibre (Dziedzic et al. 1981) or in multimode fibre (Saissy et al. 1983).

5.4.2 Optical pulse compression

The frequency chirp of a short optical pulse induced by SPM in a single-mode fibre can also be balanced in a dispersive delay line as a reflection grating pair. This technique has been applied in the compression of picosecond pulses. This method can be used to achieve the shortest pulse widths (<10 fs) in the visible spectral range (Nakatsuka et al. 1981). The optical pulse compression scheme using a fibre–grating pair assembly is shown in Figure 5.15.

Pulse compression is also possible using optical fibre gratings (Peter et al. 1994, Williams et al. 1995).

5.4.3 Soliton phenomenon

A *soliton* is a traveling pulse that propagates indefinitely without any change in shape, in its fundamental form.

For a single-mode fibre the amplitude E of the electric field satisfies (in the slowly varying amplitude approximation) the non-linear equation:

$$\mathrm{i}\left(\frac{\partial E}{\partial z} + \gamma E + k_1 \frac{\partial E}{\partial t}\right) + \frac{k_2}{2}\frac{\partial^2 E}{\partial t^2} - \frac{1}{2}\frac{n^*}{n_1} k_0 |E|^2 E = 0 \qquad (5.27)$$

where z is the longitudinal coordinate of the fibre, t the time, k_0 the propagation constant. k_1 is the inverse of the group velocity ($k_1 = \partial k/\partial \omega$); $k_2 = \lambda^2 \partial V_g/\partial \lambda$ is the group velocity dispersion ($\delta V_g/\delta \lambda < 0$), γ is the linear dumping factor, and n_1 and n^* are the linear and non-linear refractive indices (Hasegawa and Tappert 1973, Hasegawa and Kodama 1981, Doran and Blow 1983, Mollenauer et al. 1991a and 1991b, Tajima 1987).

Equation 5.27 describes the propagation of short optical pulses in a single-mode fibre, and in its lossless form ($\gamma = 0$), it is the well-known non-linear Schrödinger (NLS) equation, which supports the soliton solutions. Transforming equation 5.27 into a dimensionless form and setting $\gamma = 0$, the following equation is obtained:

$$-\mathrm{i}\frac{\partial u}{\partial z'} = \frac{1}{2}\frac{\partial^2 u}{\partial t'^2} + |u|^2 u \qquad (5.28)$$

setting:

$$t' = (t - k_1 \cdot z)\tau$$
$$z' = k_2 z/\tau^2$$
$$u = \tau E[(n^* k_0)/2n_1 |k_2|]^{-1/2}$$

τ is a normalization constant identical with time duration, u is the (complex) amplitude envelope of the pulse, z the distance along the fibre, and the time variable t is retarded time measured in a frame of reference moving along the fibre at the group velocity. The type of solution of the NLS equation obtained depends on whether the group

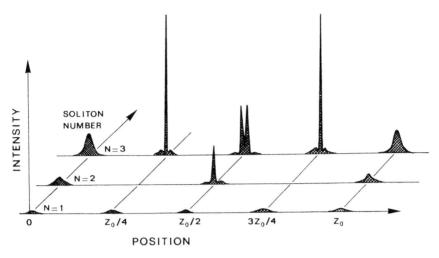

Figure 5.16. Propagation characteristics of fundamental ($N=1$) and higher order solitons $N>1$ [After Mollenauer and Stolen 1982].

velocity dispersion (GVD), $\delta V_g/\delta \lambda$, is positive or negative. For silica fibre this depends on the wavelength λ_0, a wavelength of approximately 1300 nm corresponding to zero dispersion.

In the range of negative GVD ($\delta V_g/\delta \lambda < 0$ with $V_g = \partial n_e/\partial \lambda$), the group delay of the different frequency components of the chirped pulse can be compensated by the fibre group dispersion, leading to pulse compression or to an optical soliton. If $\delta V_g/\delta \lambda > 0$, a pulse broadening occurs.

For $\lambda > \lambda_0$ (negative GVD), pulse narrowing occurs. It is also possible to observe the propagation of soliton pulses. The fundamental soliton ($N=1$) propagates without any change of shape (Zakharov and Shabat 1972) and the higher order solitons ($N>1$) have shapes that vary periodically with distance (see Figure 5.16). The pulse shapes are periodic with propagation equally efficiently as narrow broader pulses of the same energy. As seen above, the pulse narrowing and solitons result from interaction of the fibre index non-linearity with negative group velocity dispersion. It has been shown that the NLS equation has solitons as solutions for input pulses of hyperbolic secant shape with field amplitudes that are integral multiples of the amplitude of the fundamental solution. The fundamental ($N=1$) soliton $u(t)$ is obtained for a pulse of sech^2 shape:

$$u(t) = u_0 \cdot \exp(-iz/2)\operatorname{sech}(t) \tag{5.29}$$

and a peak power:

$$P_1 = \left(\frac{0.776 \lambda_0^3}{\pi^2 c n_0^*}\right) \frac{DA_e}{\tau^2}$$

τ is the pulse FWHM (full width at half maximum), D is the fibre dispersion parameter that gives the change in pulse delay with change in λ, A_e is the effective fibre core area, λ_0 the wavelength in empty space, n_0^* the index non-linearity. $P < P_1$ gives pulse broadening, $P > P_1$ gives pulse narrowing. When the sech^2 input pulse of peak power P_2 is equal to $4 \times P_1$, the first periodic behaviour occurs ($N = 2$), with a period given by:

$$Z = 0.322 \left(\frac{\pi^2 c}{\lambda_0^2}\right) \frac{\tau^2}{D} \qquad (5.30)$$

Solitons have the property that they can pass through each other and emerge with only a change in phase; they are stable with respect to small perturbations of the propagation equations. A simple explanation for the behaviour of the fundamental soliton is that the frequency chirp generated by SPM acts together with negative GVD to precisely compensate for the linear pulse dispersion.

Solitons are potentially of great interest in optical communications because they may provide a technique for data transmission at very high bit rates (Mollenauer and Smith 1988, Mollenauer et al. 1985). Soliton pulses were first described by Mollenauer et al. (1980).

A quite different type of soliton is known as a *dark soliton*. These solitons are characterized by the absence of light. They are solitons by virtue of their scattering and stability properties. However the soliton $|u| \to u_0 =$ constant when $t \to \pm\infty$ is stable and equation (5.28) has solutions in the form of localized non-linear dark excitation of the CW background. The NLS equation with $|u| = u_0$ is exactly integrable and its solution with an excitation of CW background has the general form:

$$u(z,t) = u_0 \frac{(\lambda - iv)^2 + \exp Z}{1 + \exp Z} \exp(2iu_0^2 z + i\varphi_0)$$

where

$$Z = 2vu_0(t - t_0 - 2\lambda u_0 Z) \qquad v = \sqrt{1 - \lambda^2}$$

which corresponds to the boundary conditions $|u| \to u_0$ at $t \to \pm\infty$, φ_0 and t_0 being arbitrary constants (Zakharov and Shabat 1973, Kivshar and Gredeskul 1990).

The existence of solitons was predicted theoretically by Hasegawa and Tappert (1973). Emplit et al. (1987), Krokel et al. 1988, Weiner et *al.* 1988, Kivskar and Gredeskul 1990, Barthelemy *et al.* (1985) demonstrate soliton propagation in laser beams propagating through homogeneous transparent dielectrics having a refractive index that exhibits fluctuations proportional to the local intensity.

The interplay between Kerr non-linearity and group velocity dispersion can give rise to another phenomenon known as *modulational instability*. In this process any amplitude and phase modulations of a wave travelling in the fibre may be amplified resulting in signal distortion and cross-talk.

When $\partial V_g/\partial \lambda > 0$ ($\alpha < 0$ and $\lambda < \lambda_0$ in silica fibres) pulses undergo non-linear broadening. This is the regime in which no solitons are observed. Here we see enhanced pulse broadening due to the frequency chirps developed by the SPM term reinforcing the linear dispersion of the optical fibre.

5.4.4 The soliton laser

A mode-locked laser in which the pulse width can be adjusted to any desired value down to a small fraction of a picosecond has been realized by incorporating a length of single-mode, polarization-maintaining fibre into the feedback loop of a mode-locked colour centre laser (Mollenauer and Stolen 1984) (Figure 5.17). This device (called a soliton laser) is based on the ability of single-mode fibres to support higher order solitons in the region ($\lambda > 1.3$ μm) of negative group velocity dispersion. This condition is achieved when the input pulse is four times as great as

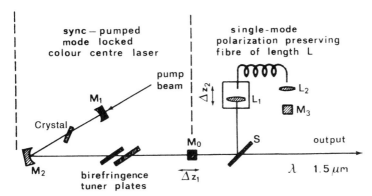

Figure 5.17. Schematic diagram of the soliton laser. Typical reflectances: $M_0 \sim 70\%$, $S \sim 50\%$; $(M_1, M_2, M_3 \sim 100\%)$. Birefringence plates: sapphire, 1 and 4 mm thick; only the thinner is used for $\tau < 0.5$ ps [After Mollenauer and Stolen 1984].

the peak power P_1 given by equation 5.29. The laser operation is thus based on the $N = 2$ soliton.

Light from the mode-locked laser is coupled through a beamsplitter and microscope objective L_1 into the polarization-maintaining fibre. At the end of the fibre a cat's-eye reflector is formed by a mirror M_3 and a lens L_2. The input fibre end and L_1 are mounted on a translation stage to facilitate adjustment (Δz_2) of the optical path length in the fibre arm so that it is an integral multiple of the main cavity length. Pulses returned from the fibre must be made coincident with those present in the main cavity. As the laser action builds up, the initially broad pulses are narrowed by passage through the fibre. These pulses are reinjected into the main cavity and compel the laser to produce narrowed pulses. The pulses in the fibre become solitons and have the same shape following the double passage though the fibre. Pulse widths of 2.0 to 0.21 picoseconds can be obtained. By compression in an external fibre, pulses shorter than 50 femtoseconds are possible.

5.5 SWITCHES

The potential of non-linearity in single-mode fibres for optical switching has been demonstrated by several authors. A review of all-optical waveguide switching in fibre and integrated optical waveguides has been presented by Stegeman and Wright (1990). Power-inducing switching has been demonstrated between the fields in a non-linear directional coupler (Ankiewicz 1988, Peng et al. 1989, Ankiewicz and Peng 1989, Chen and Snyder 1990, Leutheuser et al. 1990); between two counter-rotating circularly polarized modes linearly coupled by birefringence (Trillo et al. 1988, Daino et al. 1986, Kitayama et al. 1985); between the two linear modes of a dual-core fibre coupled through evanescent field overlap (Gusovskii et al. 1985, Maier 1984, Friberg et al. 1987, Dianov et al. 1989); and by the use of birefringent fibres (Peng and Ankiewicz 1990, Mecozzi et al. 1987). Soliton switching in non-linear fiber couplers has been investigated (Trillo et al. 1988, Dianov and Nikonova 1990, Ankiewicz and Peng 1991a,b). Ultra-fast all-optical switching using the optical Kerr effect in polarization-maintaining fibres has also been proposed (Morioka and Saruwatari 1988, Nayar and Vanherzeele 1990, Jinno and Matsumoto 1990).

5.5.1 Soliton switching in non-linear directional couplers

Solitons may be ideal for switching and signal processing applications because of their stability. Formation of solitary intense soliton pulses or trains of pulses utilizing dual-core fibres has been suggested and studied

(Trillo et al. 1988, Dianov and Nikonova 1990, Ankiewicz and Peng 1991a, b). The propagation of pulses in a non-linear dual-core coupler can be described in terms of two linearly coupled NLSEs (non-linear Schrödinger equations, 5.28). The propagation of the pulses in the moving frame is described by (Ankiewicz and Peng 1991a,b, Trillo et al. 1988, Mollenauer et al. 1980) by the equations:

$$i\frac{\partial u}{\partial z} = \pm\frac{1}{2}\frac{\partial^2 u}{\partial t^2} - Cv - |u|^2 u$$
$$i\frac{\partial v}{\partial z} = \pm\frac{1}{2}\frac{\partial^2 v}{\partial t^2} - Cu - |v|^2 v \qquad (5.31)$$

where u and v are the mode amplitudes, z and t are the normalized length and time, C is the linear coupling coefficient and '$\pm$' stands for normal or anomalous group velocity dispersion. To solve these equations, several numerical methods can be used with perturbative or exact solitary wave solutions (Romagnoli et al. 1992, Kivshar 1993).

Consider a non-linear coherent coupler made from a dual-core fibre operating in the anomalous dispersion regime. All optical switching of solitons between the two linear modes can be obtained using a frame of reference travelling at the common group velocity. Equation 5.31 is integrated by using, for example, the beam propagation method and the initial conditions:

$$u(z=0,t) = A\,sech(t/t_0) \qquad v(z=0,t) = 0$$

If $C = 0$, there is no linear coupling. Considering the anomalous dispersion regime, the soliton solutions of the isolated cores is obtained by setting $t_0 = 1$ and $A = N$, where N is an integer. The results in the anomalous and normal dispersion regimes are different. In the normal dispersion regime, the switching efficiency quickly deteriorates. In the anomalous dispersion regime, soliton switching is possible (pulse width $t_0 \approx 1$, pulse energies $2A^2 t_0 \approx 2$). Instabilities are observed when the input pulse energy $2A^2 t_0 \gg 2$ with the limit value 2 being the energy of a fundamental soliton ($N = 1$). These instabilities arise if the coupling is relatively strong ($A^2 \approx 4C \gg 1$) and also for pulses that are long compared with the soliton width ($A^2 \approx 4C = 1$ and $t_0 \gg 1$) (Trillo et al. 1988).

For the case $Z = \pi/2$ (half-beat length) and $C = 1$ the energy couples back and forth within the core for $A^2 = 1$, whereas the input pulse propagates uncoupled and undistorted for $A^2 = 4$. Figure 5.18 gives the transmission for two values between these two extreme cases.

In contrast to the quasi-CW case, where severe pulse break-up occurs, the input soliton switches all its energy between the two output cores of a fibre (for coupler length $Z = 4$), when the input

198 *Devices using non-linearities in fibres*

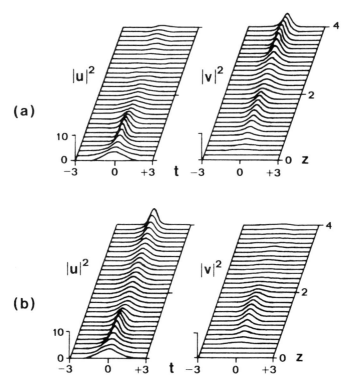

Figure 5.18. Soliton switching in a non-linear coupler excited with an input peak power: (a) $A^2 = 3.5625$, (b) $A^2 = 3.625$ [After Trillo et al. 1988].

peak power is increased by a few per cent from below to above a certain critical power. Optical switching is achieved for incident pulses with widths and peak powers comparable to those of the solitons for the uncoupled guide. By doping, for example with erbium ions, soliton switching is enhanced, although for a CW input beam the switching characteristics are degraded (Wilson et al. 1992).

In a non-linear coupler in which the arms are bent, the coupled non-linear Schrödinger equation can be reformulated; an exact solution shows that the coupling is very stable under variation of the soliton energy (Skinner et al. 1995). The propagation of ultrashort wave packets in dual-core fibres can be more completely described by a system of equations for the complex amplitudes ψ_1 and ψ_2 of the light wave envelope in the first and the second fibre.

Dianov and Nikonova (1990) investigated the possibilities provided by dual fibres for generation and filtering of ultrashort solitons.

They showed theoretically how the most intense and shortest pulse in a train of soliton pulses could be separated from a train of short pulses. They also investigated the possibilities for the generation of a train of solitons.

Dual-core fibres in the non-linear regime have bistable properties (Enns et al. 1992). They promise wide possibilities for the control of laser pulse parameters and non-linear filtering.

5.5.2 Switches using non-linear couplers

Non-linear couplers have the potential to act as fast optical switches and logic gates. As seen in Chapter 3, the coupling behaviour of a linear coupler can be described by two approaches, one being the coupling of two normal modes of individual waveguides in isolation from one another; the other being the beating of two independently propagating normal modes of the composite structure, which is more general though somewhat complicated. For weakly guiding and weakly coupling structure the two approaches give identical results.

In a non-linear coupler the oscillation in energy between the two cores depends on the initial light intensity. In analysing non-linear coupling phenomena in a coupler operating at high power, these two approaches can still be employed. When the coupling is strong, the system with composite structure has to be used. With weak coupling, the simpler normal mode approach is preferred.

The small increase in the refractive index of a material as the field increases leads to a modification of the coupled mode equation. By changing variables, the complex equations can be converted into real coupled first order differential equations (Jensen 1982, Maier 1984). The total field in the coupler can be expressed in terms of superposition of the normal modes of uncoupled waveguides 1 and 2 (see Chapter 3 equation 3.5); it can be assumed that their shapes are not significantly affected by non-linearities.

To take into account non-linear effects into the coupler, the non-linear coefficients Q_1 and Q_2 must be introduced into the coupled equations (3.6) for fields ψ_1 and ψ_2 by setting:

$$Q_1 = k_0\, n_0^* \int_{A_1} \psi_1^4\, dA \qquad Q_2 = k_0\, n_0^* \int_{A_2} \psi_2^4\, dA \qquad (5.32)$$

where n_0^* is the non-linear refractive index.

In the case of two identical fibres (identical cores) sufficiently well separated, we have:

$$C_{11} = C_{22} = 0, \quad C_{12} = C_{21} = C, \quad Q_1 = Q_2 = Q$$

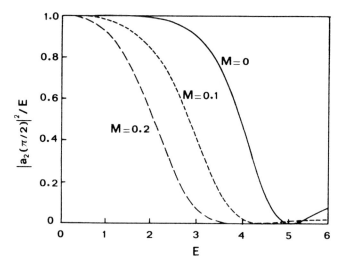

Figure 5.19. Use of non-linear coupler as an intensity-dependent switch. $E = |a_1(0)\kappa|$ [After Ankiewicz 1988].

The non-linear coupling equations are written using equation (3.6) with $b_j = a_j \exp(i\beta_j z)$:

$$-i\frac{da_1(z)}{dz} = Ca_2 + Q(1-M^2)^{-1}\left(a_1|a_1|^2 - a_2|a_2|^2 M\right)$$
$$-i\frac{da_2(z)}{dz} = Ca_1 + Q(1-M^2)^{-1}\left(a_2|a_2|^2 - a_1|a_1|^2 M\right) \quad (5.33)$$

where M is an interaction coefficient:

$$M = \int_{A_1} \psi_1 \psi_2 \, dA \quad (5.34)$$

In a linear coupler ($Q=0$) the length required for complete transfer of power is $L_b/2 (= \pi/(2C))$ (see equation (3.3) from Chapter 3). The corresponding coupling period $Z = \pi/2 = L_b/2 \cdot C$. The value of $|a_2(\pi/2)|^2 / |a_1(0)|^2$ at the length Z can be obtained for any incident power level $|a_1(0)|^2$. For a non-linear coupler, the figure 5.19 gives the values of $|a_2(\pi/2)|^2$ at this value of Z as a function of $|a_1(0)|^2$ with $|a_2(0)|^2 = 0$. At low energy, all power is transferred, but at high energy, no power is transferred; when E ($= |a_1(0)|^2$) increases by about one unit, power falls by 25–75%.

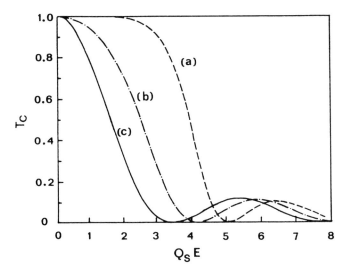

Figure 5.20. Normal-cross transmittance T_c versus Q_sE and with $Z = \pi/2$, where input power is $E = |a_1(0)|^2$ of a non-linear optical fibre coupler with (a) identical non-linearity cores $Q_r = 1$, (b) one non-linear core and one linear $Q_r = 0$, (c) cores of opposite non-linearities $Q_r = -1$. [After Peng et al. 1989].

Critical power E_c is defined as the value of E ($= |a_1(0)|^2$) which gives 50% transmittance when $\Lambda_B = \pi/(2C)$. For circular fibres in the Gaussian approximation:

$$E_c = \frac{2A_e \lambda \cdot e^{-3.2M}}{L_p \cdot n_0^* (1 - q^2)}$$

where $L_p = \pi/(2C)$ $A_e = \pi \omega_0^2$ $q = \exp(-a^2/\omega_0^2)$

r_0 is the mode spot size, a is the core radius of each fibre.

Couplers with unequal core non-linearities have advantages because the critical power is lower. Non-linear transmission depends on the way in which the non-linearities differ. Figure 5.20a shows the non-linear cross transmission T_c of a coupler with identical core non-linearities corresponding to $Q_r = Q_2 / Q_1 = 1$ and $M = 0$ (cf. Figure 5.19 for $M = 0$). Curves b and c of Figure 5.20 gives two typical cases with unequal core nonlinearities corresponding to $Q_r = 0$ and $Q_r = -1$ respectively. The larger the difference in core non-linearities, the lower is the critical power. The reductions in critical power are about 30–60%.

A non-linear bent directional coupler may act as a digital optical switch. Because of the asymptotic switching characteristic this kind of phase controlled digital switch seems to be attractive for practical applications. There is no need of an exact adjustment of the device length (Leutheuser et al. 1990).

5.5.3 Switching using birefringent fibres

The investigation of exact solutions for the evolution of the state of polarization along a non-linear single-mode birefringent fibre can be made using the Poincaré sphere representation (see Chapter 1). In the hypothesis that the field in the fibre is described by the superposition of orthogonal modes having amplitudes $A_x(z)$ and $A_y(z)$ and propagation constants β_x and β_y and in the case of the slowly varying approximation, the coupled mode equations can be written in terms of the Stokes vector **S**. This gives a system of first order ordinary differential equations

$$\frac{d\mathbf{S}}{dz} = \mathbf{V(S)}$$

It describes the motion on the Poincaré sphere of the representative point **S** subjected to the field of velocities **V(S)**, in the lossless case. In the linear case

$$\frac{d\mathbf{S}}{dz} = \mathbf{\Omega_L} \wedge \mathbf{S}$$

The sphere moves as a rigid body with angular velocity $\mathbf{\Omega_L}$ (see Figure 5.21a left). When both birefringence and Kerr effects are present, the equation is written as:

$$\frac{d\mathbf{S}}{dz} = \mathbf{\Omega} \wedge \mathbf{S} = [\mathbf{\Omega_L} + \mathbf{\Omega_{NL}(S)}] \wedge \mathbf{S}$$

The resultant motion is the sum of the two separate phenomena (see Figure 5.21b).

5.5.3.1 High birefringent fibres

A linear coherent amplifier mixer using this representation can be made (Daino et al. 1986) (see Figure 5.21). Consider a continuous pump wave, circularly polarized:

$$E_p = e_p \exp[i(\omega_p t + \phi_p)]$$

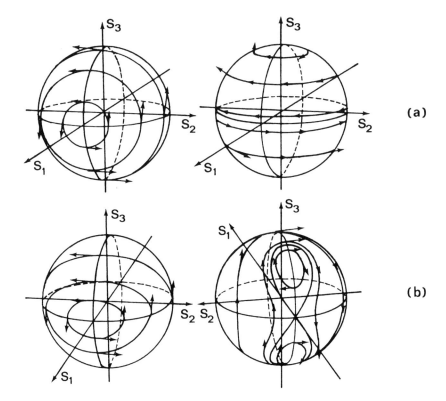

Figure 5.21. Representation of the fields of velocities associated with the notion of the state of polarization on the Poincaré sphere. Upper left: linear birefringence, with angular velocity Ω_L along S_1. Upper right: self-induced ellipse rotation (Ω_{NL}). Lower left and right: sum of the two effects, with resulting $\Omega = \Omega_L + \Omega_{\text{NL}}$. Note the presence of a separatrix at the rear of the Poincaré sphere [After Daino et al. 1986].

and a signal, with circular polarization orthogonal to the pump:

$$E_s = e_s \exp[i(\omega_s t + \phi_s)]$$

E_p and E_s are launched in the birefringent fibre of linear birefringence $\Delta\beta = \beta_x - \beta_y$. If $|e_s| \ll |e_p|$ and considering the particular case for which the amplitude of the pump is such that:

$$\Delta\beta = \beta_x - \beta_y = RP/3$$

$$P \equiv |e_p|^2 = S_1^2 + S_2^2 + S_3^2$$

P is the power in the fibre, S_1, S_2, S_3 are components of e_p. R is a function of characteristic parameters of the fibre:

$$R = 4\pi\, Z_0\, n^*/\lambda\, n_1\, A_e \tag{5.35}$$

with Z_0 the empty-space impedance and A_e the effective area of the two orthogonal modes n^* is the non-linear refractive index and n_1 the silica refractive index.

A change $\Delta S_1(0)$ around $S_1(0)$ gives a linear change $\Delta S_3(z) = G(z)\, \Delta S_1(0)$. The proportionality factor $G(z)$ depends on the length z considered and on $\Delta\beta$. At the input of the fibre the parameter S_1 oscillates around $S_1(0)(=0)$ by an amount:

$$\Delta S_1(0) = 2 e_s e_p \cos\left[(\omega_s - \omega_p) t + (\phi_s - \phi_p)\right]$$

At the output length L of the fibre, the parameter S_3 oscillates around an average value $\bar{S}_3$ by an amount $\Delta S_3(L) = G(L)\, \Delta S_1(0)$. So the signal imposes a modulation on the intensity of the pump whose depth is $G(L)$ times the one obtainable by a simple superposition of the two beams. With an increased length L an optical polarization switch is obtained, because of the relative phase relationship between signal and pump. The intensity behind a polarization splitter could then be either equal to 0 or to maximum value $|E_p|^2$ (see Figure 5.22).

5.5.3.2 Low birefringent fibres

In optical fibres with very low birefringence, non-linear polarization coupling occurs because of the ellipse rotation induced by the Kerr effect. A pure self-induced ellipse rotation appears when the linear birefringence in the fibre is negligible. It produces different features

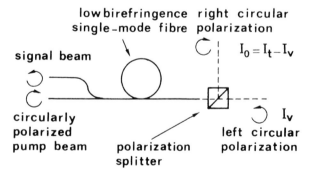

Figure 5.22. Schematic diagram of a linear coherent amplifier-mixer using a single-mode birefringent fibre [After Daino 1986].

compared with linear birefringence. The polarization coupling effect of self-induced birefringence (due to the Kerr effect) is similar to that of circular birefringence. For a fixed fibre length there is a periodic evolution of polarization as a function of the input power along a fixed fibre length. When the input power is fixed, the polarization varies in function of the fibre length. The intensity dependence of polarization coupling is not sensitive to the input polarization azimuth but to the state of polarization of the input. The field in the fibre is assumed to be:

$$E = E_x + E_y = A_x(z)e_x(x,y) + A_y(z)e_y(x,y) \tag{5.36}$$

Using the coupled mode approach, we have for the complex amplitudes A_x and A_y:

$$-i\frac{dA_x}{dz} = \beta_x A_x + \beta_+ A_x + i\beta_- A_y + R\left[|A_x|^2 + \frac{2}{3}|A_y|^2\right]A_x + \frac{R}{3}A_x^* A_y^2$$
$$-i\frac{dA_y}{dz} = \beta_y A_y + \beta_+ A_y - i\beta_- A_x + R\left[|A_y|^2 + \frac{2}{3}|A_x|^2\right]A_y + \frac{R}{2}A_y^* A_x^2 \tag{5.37}$$

with β_x, β_y the propagation constants of each polarized mode without the Kerr effect. R is given by equation 5.35. β_+ and β_- are changes of propagation constants of each mode due to polarization coupling caused by the Kerr effect.

Considering the optical Kerr effect, the solution of the coupling equations may be represented by:

$$\begin{bmatrix} A_x \\ A_y \end{bmatrix} = D \begin{bmatrix} \cos \varphi z & -\sin \varphi z \\ \sin \varphi z & \cos \varphi z \end{bmatrix} \begin{bmatrix} A_{x0} \\ A_{y0} \end{bmatrix} \tag{5.38}$$

These equations show the influence of a rotation matrix with an angle φz operating on the initial polarizations A_{x0}, A_{y0}. The term D has no influence on polarization coupling. A linearly polarized input remains linearly polarized. No cross-polarization is obtained from a circularly polarized input ($A_x = iA_y$), but an elliptically polarized input gives significant polarization coupling. Two parameters δ and θ can be introduced (Peng and Ankiewicz 1990), defined as:

$$A_{y0}/A_{x0} = \delta \exp(i\theta) \tag{5.39}$$

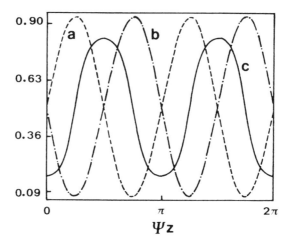

Figure 5.23. (a) $|A_x|^2/(1+\delta^2)$, (b) $|A_y|^2/(1+\delta^2)$ and (c) arg (A_x/A_y) for input light of elliptical polarization with $\delta = 1, \theta = 30°$ in equation 5.39 [After Peng and Ankiewicz 1990].

Non-linear coupling between the two linearly polarized modes occurs when: $\delta \neq 0$ and $\theta \neq \pm\frac{\pi}{2}$. Two parameters R_i and R_c are then introduced in order to deduce switching efficiency:

$$R_i = \frac{\left(|A_x|^2\right)_{\min}}{\left(|A_x|^2\right)_{\max}} = \frac{\left(|A_y|^2\right)_{\min}}{\left(|A_y|^2\right)_{\max}} \leq 1 \tag{5.40}$$

$$R_c = \frac{2\mathrm{Im}\left(A_{x0}^* A_{y0}\right)}{|A_{x0}|^2 + |A_{y0}|^2} \tag{5.41}$$

R_i measures the polarization coupling and R_c the intensity dependence relating to the polarization coupling. For optical switching, R_i can be considered as the polarization on/off ratio and R_c as a factor for power intensity. With $R_i = 0.07$ and $R_c = 0.5$, an intensity-dependent polarization 'ON' (output = 1) or 'OFF' (output 0.07) transmission is achievable (see Figure 5.23).

5.5.3.3 Twisted birefringent fibres

All-optical switching, and intensity discrimination by polarization instability in periodically twisted fibre filters, are also possible (Mecozzi et al. 1987). Periodic mode coupling introduces a power conversion between the principal axes of the twisted fibre. An input beam linearly polarized along an axis (say x) rotates its polarization by a small angle 2θ after the distance L_b (beat length). Light coupled to the orthogonal axis along successive individual coupling sections of period L_p will add up in phase if $L_p = L_b$. After a length L_c corresponding to N coupling periods, complete power transfer occurs ($2N\theta = \pi/2$). With a non-linear contribution to the refractive index, a power-dependent phase shift occurs between the linearly polarized modes of the highly birefringent fibres (Stolen and Bjorkholm 1982). The resulting beat length L'_b depends on z:

$$L'_b = L_b \{1 + R[P_x(z) - P_y(z)]/[3\Delta\beta]\}^{-1} \tag{5.42}$$

where R has the value given by Equation 5.35, P_x and P_y are the power in the x and y polarized modes and:

$$\Delta\beta = \beta_x - \beta_y = 2\pi/L_b$$

After a length L'_b, a beam polarized along the x-axis will couple to the y-axis with a power $\sin^2(2\theta')$ with $\theta' < \theta$. Full polarization is achieved after N' coupling sections ($N' > N$). The non-linear coupled equations for the complex amplitudes A_x and A_y of the linear modes polarized in the x and y directions along a periodically twisted fibre are similar to equation (5.37) (Mecozzi et al. 1987).

The local linear birefringence at a distance z is now given by:

$$\Delta\beta \left[1 + \varepsilon^2 \cos^2\left(\frac{2\pi}{L_b}z + \phi\right)\right]^{1/2} \approx \Delta\beta \tag{5.43}$$

since $\varepsilon (= 4C/\Delta\beta) \ll 1$ where C is the linear coupling coefficient, and the principal axes periodically rotate along z by a small angular amplitude $\theta = \varepsilon$, ϕ is an arbitrary phase.

When the input power P_0 is equal to a critical value P_c, the initial mismatch is so high that its sign no longer reverses and the input beam splits between the axes ($P_x = P_y$). This is a spatially unstable state (Matera and Wabnitz 1986, Jensen 1982, Daino et al. 1986, Winful 1986, Trillo et al. 1986, Wabnitz et al. 1986). When P_0 becomes greater than P_c the polarization state along the fibre always has a dominant x-component. Thus, as seen in Figure 5.24, the polarization state at the output of a length L_b' can be switched from the y- to the x-axis through a slight increase in P_0 across P_c.

5.5.3.4 Birefringent fibres with cross axis

To fully utilize the long interaction characteristics, linear pump polarization should be maintained by using high-birefringent polariza-

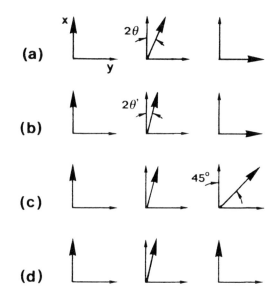

Figure 5.24. Schematic evolution of an input wave linearly polarized along an axis: (a) Linear conditions after a distance L_b and after several coupling sections, (b) Power $p = 2P_0/P_c < 2$ after distances L_b' and N' sections, (c) Same as (b), with $p = 2$, (d) Same as (b), with $p > 2$ [After Mecozzi et al. 1987].

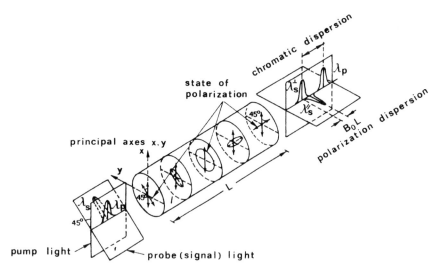

Figure 5.25. Principle of the optical Kerr switching operation in PM–SM fibre [After Morioka and Saruwatari 1988].

tion-maintaining (PM) fibres as the Kerr medium. The temperature dependent fluctuations of signal pulses make the switching unstable, and the polarization dispersion degrades the switching bandwidth. The method of compensating for these fluctuations is to link two identical fibres, with the fast axes crossed at 90° (Dziedzic et al. 1981, Morioka and Saruwatari 1988) (see Figure 5.25).

Pump and signal pulses are coupled and co-propagating into the first fibre with 0° and 45° linear polarizations with respect to one of the optical axes of the fibre. The pump power modifies the refractive index and the state of polarization at the end of the fibre can be set at 135° linear polarization. Between the two orthogonal directions x and y, the phase difference $\Delta\Phi(t)$ along a length L is given by:

$$\Delta\Phi(t) = \frac{2\pi}{\lambda_0} \Delta n(t) L + \frac{2\pi}{\lambda_0} B_0 L \tag{5.44}$$

$\Delta n(t)$ is the intensity-induced refractive index change by Kerr effect, B_0 is the modal birefringence of the fibre and λ_0 the wavelength. $\Delta n(t)$ is the refractive index difference between the two components:

$$\Delta n(t) = \Delta n_x(t) - \Delta n_y(t) \tag{5.45}$$

$n_x(t)$ and $n_y(t)$ are depending on the pump intensity profile. If the pump intensity $I_p(t)$ is chosen to be the x polarization and if dispersions are negligible as compared to the pump and pulse durations, then $\Delta n(t) \cong n_0^* I_p(t)$. n_0^* is the optical Kerr constant (see equation 5.26) in which the loss due to dispersion is not taken into account. The probe pulse is split into two orthogonal polarizations. A modal birefringence around 3×10^{-4} separates the two polarization components by as much as 1 ns km^{-1} (10 ps per 10 m). A change in B_0, due to temperature change, causes fluctuations in the term $\Delta\Phi' = (2\pi/\lambda_0)B_0 L$ which can reach several radians per degree celsius. Compensation for fibre birefringence (while keeping its polarization-maintaining properties) is then achieved by splicing the second identical fibre with its fast axis crossed at 90°.

In the absence of the control pulses, the analyser blocks the signal pulses. In the presence of the control pulses the output polarization state of the signal changes due to the intensity induced by birefringence change. This technique makes ultrafast all-optical switching possible.

Dual wavelength operation is generally considered unattractive due to the losses associated with group velocity dispersion (GVD). However, it can generate the uniform phase required for complete switching. Nayar and Vanherzeele (1990) have demonstrated experimentally that GVD can be exploited to achieve complete optical switching for a range of temporal delays, so that the fast pulse (the longer of the two wavelengths) can pass through the peak of the slow pulse. Complete optical switching of 2–30 ps signal pulses in the 0.83–1.0 μm wavelength range has been demonstrated using 50 ps control pulses at 1.053 μm. This is achieved using an initial temporal offset between the pulses and group velocity dispersion to enable the fast pulse to pass through the slow one. Minimum switching power results when the initial offset is one half of the relative group delay.

5.5.4 Switching using non-linear fibre loop mirror

5.5.4.1 With non-birefringent fibre

A non-linear fibre loop mirror (NOLM) is a fibre Sagnac interferometer consisting of a fibre that connects the input ports of an X coupler, forming a closed loop. If a 50/50 split coupler is used to balance the Sagnac interferometer, injection of a wave gives (theoretically) a null output. If the power coupling ratio is no more equal to 1/2 and using an intense data pulse into the loop, a non-linear phase shift occurs, and the propagation of the light will no longer be identical for the two paths, because the intensity is different. The phase velocity is intensity dependent bias the non-linear refractive index and the interferometer is not balanced (Stegemam et Wright 1990, Doran and Wood 1988, Shi 1994, Mahgereftech et Chbat 1995, Olsson et Andrekson 1995, Doran and Wood 1988).

For a wavelength λ and a power coupling ratio CR, the output power P_{out} at fibre length L is given by:

$$P_{\text{out}} = P_{\text{in}} \left[1 - 2CR(1 - CR)\{1 + \cos[(1 - 2CR)\Delta\phi]\} \right]$$

with:

$$\Delta\phi = \frac{4\pi n^*}{\lambda A_e} L P_{\text{in}}$$

n^* is the non-linear Kerr coefficient in m^2/V^2 (see equation 5.26) and A_e is the effective core area. The output power will exhibit maxima and minima according to the input peak power.

This device can be used for ultrafast switching (Lima and Sombra 1999, Chan et al. 1998), for soliton compression (Chusseau et al. 1994), noise filtering (Olsson and Andrekson 1995), high-speed wavelength conversion of digital data (Mahgereftech and Chbat 1995) and transmission (Boscolo et al. 2000).

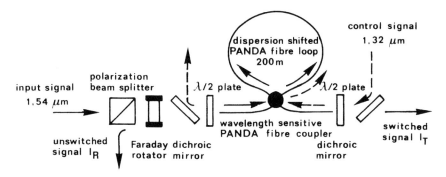

Figure 5.26. Schematic configuration of the non-linear Sagnac interferometer switch [After Jinno and Matsumoto 1990].

5.5.4.2 With PANDA fibre

An ultrafast, low power, highly stable all-optical switching is possible using a polarization-maintaining fibre into a non-linear Sagnac interferometer (Jinno and Matsumoto 1990). The key component is a dispersion-shifted PANDA fibre loop (see Figure 5.26). It has an equal group velocity at the wavelengths of the input signal and the control pulse. To achieve low power, high stability, and loss free switching, a small core area dispersion-shifted polarization-maintaining fibre loop (200 m in length) is used. A wavelength-sensitive polarization-maintaining fibre coupler, with a power coupling ratio of nearly 50% at the wavelength of the input signal and 0% at the wavelength of the control pulse is employed. A weak input signal, to avoid the non-linear effect induced by the signal itself, is launched into a port of the wavelength sensitive polarization maintaining fibre coupler through a 45° Faraday rotator. It is split into two counter-propagating signals in the polarization-maintaining fibre loop. A high-intensity control pulse train is launched into the other port of the coupler and propagated in an anti-clockwise direction.

In the absence of the control signal, the two counter-propagating signals return in phase through the exact same path length to the coupler. As seen in Chapter 3 and 4 they recombine and are fully reflected. This reflected light, passing through the 45° Faraday rotator twice, is orthogonally polarized with respect to the input signal. So the polarization beamsplitter extracts it.

In the presence of the control signal, the induced cross-phase modulations break the equality between the two counter-propagating paths, and some part of the input signal is transmitted.

By this technique, highly stable and loss-free all-optical switching has been demonstrated at more than 5 Gbs^{-1}.

5.6 NON-LINEAR FIBRE INTERFEROMETER

A Mach-Zehnder interferometer with a non-linear path can also been used (Imoto *et al.* 1987, Crosignagni *et al.* 1986). The multistability of interferometers with non-linear elements has been studied. A device based on an interferometer with common mode compensation (Backman 1989) but using a fibre with an intensity-dependent refractive index has been investigated (Babkina *et al.* 1990). An analytical description of the device with a non-linear fibre is presented and the possibility of multistable regimes is discussed. The advantages of using solitons for all-optical switching in non-linear interferometers have also been discussed by Doran and Wood (1988). In an interferometer, two-soliton solutions of the non-linear scalar Schrödinger equation (NLSE) are propagated uncoupled. Switching is obtained through the differential phase shifts acquired by the two solitons in the two arms through a phase-sensitive splitter such as an X-coupler.

5.7 MODULATOR AND LOGIC GATE

Non-linear refraction and absorption can be used to obtain many of the desired logic functions in a single-mode fibre system by modulating the fibre-to-fibre coupling factor without resonators or feedback (Jeong and Marhic 1991, Normandin 1986). Light from the single-mode input fibre 1 is transmitted through a thin silicon wafer (non-linear material) to the output fibre 2 (see Figure 5.27). Light from the single-mode gate fibre 3 is incident upon the wafer and creates electron-hole pairs that modify the refractive index of the medium. Since light emitted by a single coupled-mode fibre has a spatially Gaussian distribution, a lensing effect, which modifies the effective numerical aperture of the input fibre as seen by the output fibre, is induced. The resulting change in refractive index Δn, at the peak of the pulse is given by:

$$\Delta n_1 = -\frac{e^2 \tau \alpha(T) \sqrt{\pi}}{4 n_1 m_{\text{eh}} \omega^2 \varepsilon_0 h\nu} + \frac{\alpha(T) \sigma_{\text{eh}}(T) \tau^2 \pi}{8 C' h\nu} \frac{\partial n_1}{\partial T} I_0^2 \qquad (5.46)$$

where e is the electron charge, m_{eh} the optical effective mass of an electron-hole plasma, $\alpha(T)$ the linear absorption coefficient and $\sigma_{\text{eh}}(T)$ the free carrier absorption coefficient, depending on the temperature T, n_1 is the refractive index of the wafer for incident light of frequency ν, C' the heat capacity, τ the Gaussian pulse width given by:

$$I = I_0 \exp(-t^2/\tau^2) \qquad (5.47)$$

The photo-generated carrier density and the lattice temperature are assumed constant throughout the thickness of the wafer.

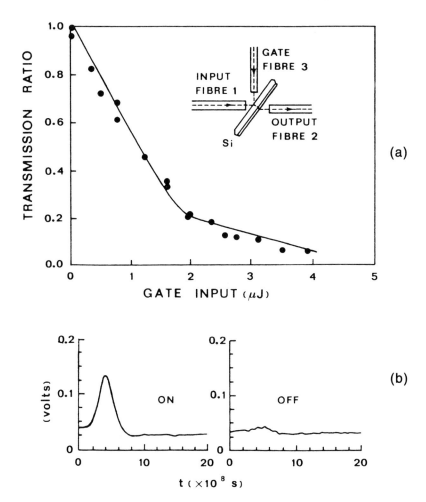

Figure 5.27. (a) Transmission ratio versus gate fibre (3) energies for a 63 μm Si wafer, (b) typical laser pulse in ON and OFF states (4 μJ applied on gate fibre (3)). [After Normandin 1986].

Equation 5.46 may be used to calculate the effective focal length f of the negative lens thus created at low energies:

$$f \cong a_0^2/2d\Delta n_1$$

a_0 is the Gaussian spot size of the fibres and d the wafer thickness. The usual lens formula relates the magnification, position and effective numerical aperture of the input fibre as seen by the output fibre.

214   *Devices using non-linearities in fibres*

The typical pulse in the ON and OFF states, shown in Figure 5.27(b), has been obtained from a single longitudinal mode passively Q-switched Nd: YAG laser.

All-optical logic gates performing Boolean functions (ON, AND, OR XOR, NAND, NOR, INV) have been proposed and demonstrated using non-linear interferometers (Jeong and Marhic 1991). The basic device is a non-linear Mach-Zehnder interferometer based on the cross-phase modulation between a signal and a pump with two different wavelengths. To overcome the instability of this interferometer due to thermal and acoustic fluctuations a Sagnac interferometer has also been used.

AND gates have been realized using soliton phenomenon with on/off contrast ratios greater than 20:1 (Islam *et al.* 1992).

5.8 OPTICAL FIBRE TRANSISTOR

An electronic transistor can amplify a small signal, which is added to a d.c. bias condition. This concept is also applicable to an all-optical system, but the mechanism is, of course, quite different. A non-linear optical coupler can function as an optical transistor when operating near to its critical power level (Ankiewicz 1989, Maier 1987, Gusovskii *et al.* 1985). At low powers, a coupler made from two parallel optical waveguides, labelled 1 and 2, works as a linear coupler. Power swamps between the guides 1 and 2 when non-linear effects appear at high powers. If high power is launched into guide 1, then a very small part is coupled into guide 2. At an intermediate power there is a critical level, where the power in guide 1 decreases proportionally with coupler length.

If the input intensity is set at a level roughly equal to the critical power a small optical signal superimposed on this can make an optical transistor. The input is a sequence of very short pulses (L10 ps) of stable constant amplitude. By using these short pulses, difficulties arising from thermal effects and from Brillouin scattering are removed. The output pulses are deeply modulated by the signal thus constituting amplification. Wavelength longer than the pump wavelength is introduced by SRS and is filtered out using a grating. Gurashi *et al.* (1987) have demonstrated experimentally the general feasibility of the optical transistor. A small gain of 3–5 has been obtained using a dual-core fibre guide.

An analysis of the device is given by Ankiewicz (1989). In a transistor made from identical fibres with $M = 0$ (as defined in equation 5.34) i.e., large separation between the cores, the amplification A takes the value:

$$A \cong \frac{4 + \exp(2L)}{8}$$

where $L(>3)$ is the normalized distance ($L = CL_p$), C is the coupling coefficient and L_p the actual length along the coupler. Thus A increases exponentially with the length of the transistor.

REFERENCES FOR CHAPTER 5

Aoki Y 1988 *J. of Lightwave Technol.* **6** 1225–39
Aoki Y 1989 *Opt. Quant. Elect.* **21** 89–104
Aoki Y, Kishida S and Washio K 1986 *Appl. Optics* **25** 1056–60
Aoki Y, Kishida S, Honmou H, Washio K and Sugimoto M 1983 *Electron. Lett.* **19** 620–2
Ankiewicz A 1988 *Optical and Quantum Electron.* **20** 329–37
Ankiewicz A 1989 *IEE Proc. Part J.* **136** 111–17
Ankiewicz A and Peng GD 1989 *Opt. Comm.* **73** 75–80
Ankiewicz A and Peng GD 1991a *Opt. Comm.* **84** 71–5
Ankiewicz A and Peng GD 1991b *Intern J. of optoelectron* **6** 15–22
Auyeung J and Yariv A 1978 *IEEE J. of Quantum electron.* **14** 347–52
Babkina TV, Bass FG, Bulgakov SA, Grigor'yants VV and Konotop VV 1990 *Opt. Comm.* **78** 398–402
Backman A B 1989 *J. Lightwave Technol.* **7** 151
Barthelemy A, Maneuf S and Froehly C 1985 *Opt. Comm.* **55** 201–6
Bayvel P and Giles IP 1990 *Optics Comm.* **75** 57–62
Bayvel P, Giles IP and Radmore PM 1989 *Opt. Quant. Elect.* **21** 113–28
Blow KJ and Doran NJ 1987 *IEE Proc. PtJ* **134** 138–44
Boscolo S, Nijhof JHB and Tutsym S 2000 *Opt. Lett.* **25** 1240–42
Botineau J, Leycuras C and Montes C 1988 *SPIE* **963** 132–7
Botineau J, Leycuras C, Montes C and Picholle E 1989 *J. Opt. Soc. Am. B.* **6** 300–12
Botineau J, Leycuras C, Montes C and Picholle E 1994 *Annales des Télécommunications* **49** 479–89
Chan CK, Chen L K and Cheung KW 1998 *J. Opt. Comm.* **19** 67–71
Chraplyvy AR 1990 *IEEE J. Lightwave Technol.* **8** 1548–57
Chraplyvy AR and Tkach RW 1986 *Electron. Lett.* **22** 1084–5
Chen Y and Snyder AW 1990 *IEEE J. of Lightwave Technol.* **8** 802–10
Crosignani B 1992 *Fiber and Integrated Optics* **11** 235–52
Crosignani B, Diano B, Diporto P and Wabnitz S 1986 *Opt. Comm.* **59** 309
Cotter D 1987 *Optical and Quantum Electron.* **19** 1–17
Cotter D 1982 *Electron. Lett.* **18** 495–6
Daino B, Gregori G and Wabnitz S 1986 *Opt. Lett.* **11** 42–4
Dask ML and Melman P 1985 *IEEE J. of Lightwave Technol.* **3** 806–13
Davey ST, Williams DL, Ainslie BJ, Rothwell WJM and Wakefield B 1989 *IEE Proc. PtJ.* **136** 301–6

Desurvire E, Digonnet MJF and Shaw HJ 1985 *Opt. Lett.* **10** 83–5
Desurvire E, Digonnet MJF and Shaw HJ 1986 *IEEE J. Lightwave Technol.* LT **4** 426–43
Desurvire E, Imamoglu A and Shaw HJ 1987 *IEEE J. of Lightwave Technol.* **5** 89–96
Dianov EM and Nikonova ZS 1990 *Opt. and Quantum Electron.* **22** 427–31
Dianov EM, Kuznetsov AV, Maier AA, Okhotnikov OG, Sitarsky KY and Shcherbakov IA 1989 *Optics Comm.* **74** 152–4
Doran NJ and Blow J 1983 *IEEE J. Quantum Electron.* **19** 1883
Doran NJ and Wood D 1988 *Opt. Lett.* **13** 56–8
Dziedzic JM, Stolen RH and Ashkin A 1981 *Appl. Opt.* **20** 1403–6
Emplit P, Hamaide JP, Reynaud F, Froehly C and Barthelemy A 1987 *Optics Comm.* **62** 374–9
Enns RH, Edmundson DE, Rangnekar SS and Kaplan CE 1992 *Optical and Quantum Electronics* **24** 1295–314
Foley B, Dakss ML, Davies RW and Melman P 1989 *J. of Lightwave Technol.* **7** 2024–32
Friberg SR, Silberberg Y, Olivier MK, Andrejco MJ, Saifi MA and Smith P M 1987 *Appl. Phys. Lett.* **51** 1135
Gusovskii DD, Dianov EM, Maier AA, Neustruev VB, Shklovskii EI and Shcherbakov IA 1985 *Sov. J. Quant. Electron.* **15** 1523
Hasegawa A and Kodama Y 1981 *Proc. IEE Pt* **69** 1145
Hasegawa A and Tappert F 1973 *Appl. Phys. Lett.* **23** 142–4
Henry WM 1992 *Intern. J. of Optoelectronics* **7** 453–78
Hill KO, Johnson DC, Kawasaki BS and MacDonald RI 1978 *J. Appl. Phys.* **49** 5098–106
Hirose A, Takushima Y and Okoshi T 1991 *J. Opt Commun.* **12** 82–5
Ippen EP and Stolen RH 1972 *Appl. Phys. Lett.* **21** 539–41
Imoto N, Watkins S and Sasaki Y 1987 *Optics Comm.* **61** 159
Inoue K 1994 *IEEE Photonics Technol. Lett.* **6** 1451–3
Inoue K and Toba H 1992 *IEEE Photonics Technol. Lett.* **4** 69–72
Inoue K and Toba H 1995 *IEEE J. of Lightwave Technol.* **13** 88–93
Islam MN, Soccolich CE and Gordon JP 1992 *Opt. and Quantum Electron.* **24** 1215–35
Jensen SM 1982 *IEEE J. Quantum Electron.* **QE 18** 1580–3
Jeong JM and Marhic ME 1991 *Optics Commun.* **85** 430–6
Jinno M and Matsumoto T 1990 *IEEE Photonics Technol. Lett.* **2** 349–51
Kean PN, Smith K and Sibbett W 1987 *IEE Proc. PtJ* **134** 163–70
Kikuchi K and Lorattanasane C 1994 *IEEE Photonics Technol. Lett.* **6** 992–4
Kitayama K, Kimura Y and Seikai S 1985 *Appl. Phys. Lett* **46** 317
Kivshar YS 1993 *Opt. Lett.* **18** 7–9

Kivshar YS and Gredeskul SA 1990 *Optics Comm.* **79** 285–90
Krokel D, Halas NJ, Guiliani G and Grishowsly 1988 *Phys. Rev. Lett.* **60** 29
Kuckartz M, Schulz R and Harde H 1987 *Opt. and Quantum Electron.* **19** 237–46
Lee HY, Wu J, Kao MS and Tsao HW 1992a *IEE Proc. J* **139** 272–278
Lee HY, Tsao HW, Kao MS and Wu J, 1992b *J. of Optical Communication* **13** 99–103
Leutheuser V, Langbein U and Lederer F 1990 *Opt. Comm.* **75** 251–5
Lima J L S and Sombra ASB 1999 *J. Opt. Comm.* **20** 82–7
Lin C 1986 *IEEE J. Lightwave Technol.* **4** 1103–15
Liu QD, Chen JT, Wang QZ, Ho PP and Alfano RR 1995 *IEEE Photonics Technol. Lett.* **7** 517–19
Maeda MW, Sessa WB, Way WI, Yi-Yan A, Curtis L,, Spicer R and Laming RI 1990 *IEEE J. of Lightwave Technol.* **81** 1402–8
Mahgerefteh D and Chbat MW 1995 *IEEE Photonics Technol. Lett.* **7** 497–9
Maier AA 1984 *Sov. J. Quantum Electron.* **14** 101
Maier AA 1987 *Sov. J. Quantum Electron.* **17** 1013–17
Matera F and Wabnitz S 1986 *Opt. Lett.* **11** 467
Mecozzi A, Trillo S, Wabnitz S and Daino B 1987 *Opt. Lett.* **12** 275–7
Mochizuki K, Edagawa N and Iwamoto Y 1986 *J. Lightwave Technol.* **4** 1328–33
Mollenauer LF and Smith K 1988 *Opt. Lett.* **13** 675–7
Mollenauer LF and Stolen RH 1982 *Fiber Opt. Techn.* **4** 193
Mollenauer LF and Stolen RH 1984 *Opt. Lett.* **9** 13–15
Mollenauer LF, Evangelides SG and Haus HA 1991a *IEEE J. Lightwave Technol.* **9** 194–7
Mollenauer LF, Stolen RH and Gordon JP 1980 *Phys. Rev. Lett.* **45** 1095
Mollenauer L F, Evangelides SG and Gordon JP 1991b *IEEE J. Lightwave Technol.* **9** 362–7
Mollenauer LF, Stolen RH and Islam MN 1985 *Opt. Lett.* **10** 229–31
Montes C, Legrand O, Rubenchik AM and Relke IV 1989 *Intern. Workshop on Non-linear and Turbulent Process in Physics* **2** 1250–66
Morioka T and Saruwatari M 1988 *IEEE J. Select Areas Commun.* **6** 1186–98
Morioka T, Saruwatari M and Takada A 1987 *Electron. Lett.* **23** 453–4
Nakamura K, Kimura M, Yoshida S, Hidaka T and Mitsuhashi Y 1984 *J. Lightwave Technol.* **2** 379–81
Nakashima T, Seikai S and Nakazawa M 1986 *J. Lightwave Technol.* **4** 569–73
Nakatsuka H, Gricheowky D and Balant AC 1981 *Phys. Rev. Lett.* **47** 910

Nayar BK and Vanherzeele H 1990 *IEEE Photonic Technol. Lett.* **2** 603–5

Normandin R 1986 *Opt. Lett.* **11** 751–3

Ovsyannikov DV, Kuzin EA, Petrov MP and Belotitskii VI 1991 *Opt. Comm.* **82** 80–2

Olsson NA and Van der Ziel JP 1986 *Electron. Lett.* **22** 488–90

Olsson BE and Andrekson PA 1995 *J. of Lightwave Technol.* **13** 213–15

Peng GD and Ankiewicz A 1990 *Opt. Quantum Electron* **22** 343–50

Peng GD, Ankiewicz A and Snyder AW 1989 *Intern. J. of Opto Electon.* **4** 389–96

Peter DS, Hodel W and Weber HP 1994 *Opt. Communication* **112** 59–66

Picholle E, Montes C, Leycuras C, Legrand O and Botineau J 1991 *Phys. Rev. Lett.* **66** 1454–7

Pocholle JP, Papuchon M, Raffy J and Desurvire E 1990 *Rev. Techn. Thomson CSF* **22** 187–268

Pocholle JP, Raffy J, Papuchon M and Desurvire E 1985 *Opt. Eng.* **24** 600–8

Reis A, Vermelho M, Nicacio D, Gouveia E and Gouveia-Neto A 1995 *Fiber and Integrated Optics* **14** 179–92

Romagnoli M, Trillo S and Wabnitz S 1992 *Optical and Quantum Electronics* **24** S-1237–67

Sassy A, Botineau J and Ostrowsky DB 1983 *Appl. Opt.* **22** 3869–73

Shi CX 1994 *Optics Commun.* **107** 276–80

Shibata N, Nosu K, Iwashita K and Azuma Y 1990 *IEEE J. on Selected Area in Comm.* **8** 1068–77

Skinner IM, Peng GD, Malomed BA and Shu PL 1995 *Optics Communications* **113** 493–7

Smith RG 1972 *Appl. Opt.* **11** 2489–94

Spirit DM, Blank LC, Davey ST and Williams DL 1990 *IEE Proc. PtJ* **137** 221–4

Stegeman GI and Wright EM 1990 *Opt. and Quantum Electron.* **22** 95–122

Stokes LF, Chodorow M and Shaw HJ 1982 *Opt. Lett.* **7** 509–11

Stolen RH 1979 *IEEE J. Quantum Electron.* **15** 1157–60

Stolen RH and Bjorkholm JE 1982 *IEEE J. of Quantum Electron.* QE **18** 1062–72

Stolen RH and Lin C 1978 *Phys Rev. A* **17** 1448–53

Stolen RH, Bosch MA and Lin C 1981 *Optic Lett.* **6** 213

Stolen RH, Lee C and Jain RK 1984 *J. Opt. Soc. Am.* **1** 652–7

Sudo S and Itoh H 1990 *Opt. and Quantum Electron.* **22** 187–212

Tajima K 1987 *Opt. Lett.* **12** 54–6

Tanaka Y and Hotate K 1995 *IEEE Photonics Technol. Lett.* **7** 482–84

Thomas PJ, Van Driel HM and Stegeman GIA 1980 *Appl. Opt.* **19** 1906–8

Tkach RW and Chraplyvy AR 1989 *Opt. Quant. Electron.* **21** 105–12

Tkach RW, Chraplyvy AR, Forghieri F, Gnauck AM and Derosier RM 1995 *IEEE J. of Lightwave Technol.* **13** 841–9

Tomita A 1983 *Opt. Lett.* **8** 412 14

Trillo S, Wabnitz S, Stolen RH, Assanto G, Seaton CT and Stegeman GI 1986 *Appl. Phys. Lett.* **49** 1224

Trillo S, Wabnitz S, Wright EM and Stegeman GI 1988 *Opt. Lett.* **13** 672–4

Veith G 1988 *Fiber and Integrated Optics* **7** 205–15

Wabnitz S, Wright EM, Seaton CT and Stegeman GI 1986 *Appl. Phys. Lett.* **49** 838–40

Weiner AM, Heritage JP, Hawkins RJ, Thurson RN and Kirschner FM, Leaird DE and Tomlinson WJ 1988 *Phys. Rev. Lett.* **61** 2445

Williams JAR, Bennion I and Zhang L 1995 *IEEE Photonics Technol. Lett.* **7** 491–3

Wilson J, Stegman GI and Wright EM 1992 *Opt. and Quantum Electronics* **24** 1325–36

Winful HG 1985 *Appl. Phys. Lett.* **46** 527

Yu J and Jeppesen 2000 *Opt. Lett.* **25** 393–5

Zakharov VE and Shabat AB 1972 *Soviet Phys. JET* **34** 62

Zakharov VE and Shabat AB 1973 *Soviet Phys JET* **37** 823

EXERCISES

5.1 – A power $P(0) = 2$ W is injected in a very long single-mode polarization maintaining fibre with a core radius $a = 10$ μm.

a–What is the length L_e for which power pump level equals the losses?

b–Calculate the Raman gain G[dB] if the Raman gain coefficient is $g_r = 9 \times 10^{-14}$ m/W and $\alpha_p = 0.05$ dB/km.

5.2 – Calculate from equation 5.16 the acoustic frequency ν_B created by a pump signal at $\lambda_P = 1.3$ μm. By application of phase matching condition, determine the Stokes backward signal frequency and verify the obtained value with energy conservation formula.

5.3 – Using the acoustic frequency ν_B created by pump signal at $\lambda_P = 1.55$ μm.

Calculate the Brillouin linewith $\Delta\nu_B$.

Calculate the Brillouin gain coefficient g_B ($n = 1.46$, $p_{12} = 0.283$, $K = 1, c = 3 \times 10^8$ m/s, $\rho = 2210$ kg m^{-3}, $V_A = 6000$ m s^{-1}).

5.4 – Knowing that the silica chromatic dispersion is $D_c = -\dfrac{\lambda}{c}\dfrac{d^2 n}{d\lambda^2} = 1$ ps/nm/km at $\lambda_p = 1.32$ μm, what is the group velocity dispersion $D(\lambda)$

necessary to have phase matching for four wave mixing and what is the corresponding bulk dispersion for a frequency shift corresponding to $\lambda_s = 1.67 \ \mu m$?

5.5–At which values of fibre length z a pulse is becoming identical to itself if it has an initial shape $\text{sech}^2 t/\tau$ at $z_0 = 0$ and a pulse width $\tau = 50$ or 30 ps for $\lambda_0 = 1.65 \ \mu m$ and $D = 1$ ps/km/nm?

Chapter 6

LASERS AND AMPLIFIERS BASED ON RARE-EARTH DOPED FIBRES

In this chapter, applications of fibre optical devices studied in previous chapters are described as the main components of new devices used in communication and in sensors, amplifiers and lasers.

Direct optical amplification of signal light is of considerable interest, especially in optical communication systems and it can be achieved by semiconductors or in optical fibres. In the latter, as seen in Chapter 5, it can be achieved by using non-linear phenomena such as stimulated Raman scattering (SRS), stimulated Brillouin scattering (SBS) or stimulated four-photon mixing (SFPM). Raman amplifiers have wide bandwidth, low loss outside the gain bandwidth, ultralow noise and high saturation power. Gains higher than 30 dB are achievable. However, they require very high pump powers. They are used in telecommunication systems as preamplifiers.

Brillouin amplifiers also provide high gains, in the range of 50 dB for pump powers available from DFB lasers. Their bandwidth ranges from 15 MHz to 200 MHz depending on the pumping scheme. The narrow bandwidth has been exploited for carrier regeneration in homodyne detection and the higher bandwidth scheme for channel selection.

Rare-earth doped single-mode fibre amplifiers have also found many applications in optical communications and sensors, and are the most promising for future development. An increasing number of laboratories have become involved in research and development of optical devices (amplifiers and lasers) based on rare-earth doped fibre amplifiers (REDFAs). Many papers and books are devoted to the subject (e.g., Artiglia *et al.* 1994, Urquhart 1989, Payne 1992). A detailed review of the state of the art in this important area of REDFL (rare-earth doped fibre lasers) and REDFAs is given by J.F. Digonnet 1993. The broad fluorescence linewidth of rare-earth ions in glass may permit the construction of tunable sources and broad-band amplifiers. Rare-earth doped fibre amplifiers are expected to become key components in long distance, high-speed optical fibre communication.

Erbium doped fibre amplifiers (EDFAs) show great potential in light wave communications systems at 1.56 μm, as demonstrated by Laming *et al.* 1989, Ainslie 1991. The characteristics are favourable for use in high-speed fibre optics communication systems: high gain, travelling wave amplification in the low loss 1.5 μm window, low insertion losses polarization-independent gain, low noise and immunity from cross-talk. EDFAs are being used in various light wave system experiments for long distance high-speed transmission.

Using several *EDFAs* in line and in dispersion-shifted single-mode fibres, it is possible to transmit through lengths of fibre greater than 1100 km. Soliton transmission in single-pass long fibre with EDFAs has been demonstrated over 100–150 km with frequencies of 3–10 Gb s^{-1}.

Power EDFAs have been used in trunk and distribution network applications, in multichannel amplification and in multichannel amplitude modulated cable television transmission. Fibre amplifiers may also be useful for simultaneous amplification and compression of optical pulses. Sub-picosecond pulse compression (to less than 300 fs) is obtained and receivers using erbium doped fibre preamplifiers can have a very good performance.

There are also many potential applications for REDFLs. Fibre loop reflectors have been applied to doped fibre lasers, and tunable doped single-mode fibre lasers have been constructed. Demonstrations of low threshold laser optical sources using ring fibre resonators have been reported in the literature and dye ring lasers have been made in single-mode fibre versions by means of evanescent field couplers.

6.1 RARE-EARTH DOPED FIBRE AMPLIFIERS (REDFAs)

REDFAs are especially promising because high gains can be achieved; also, they can be pumped by a laser diode. Moreover, the noise figure of erbium doped fibre laser amplifiers is expected to be close to the theoretical 3 dB limit.

Operation of a fibre optics amplifier requires a pump and a signal to be injected collinearly into a fibre. The two guided waves may travel in the same or opposite directions. Energy absorbed from the pump beam provides the population inversion necessary for amplification (Figure 6.1).

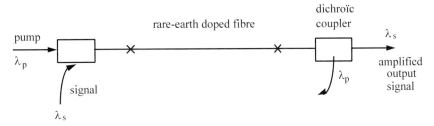

Figure 6.1. Schematic diagram of a rare-earth doped fibre amplifier (REDFAs).

6.1.1 Principle

In solid-state devices material, the trivalent (3+) level of ionisation is the most stable for rare-earth ions (lanthanides), and most optical devices use trivalent ions. The observed infrared and visible optical spectra of these trivalent ions is a consequence of transitions between $4f$ states. The lanthanides are characterised by the filling of the $4f$ shell while the external shells are already filled. The first of the lanthanide series is cerium (Ce) and the last is lutetium (Lu). The schemes of the energy levels can be calculated from the electronic structure. Several examples are given as on Figures 6.2 and 6.3. (The Russell-Saunders

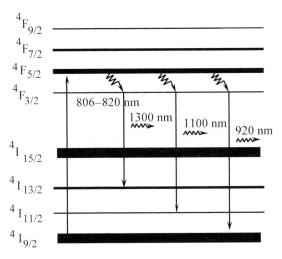

Figure 6.2. Energy level diagram for the trivalent neodymium ion Nd^{3+}.

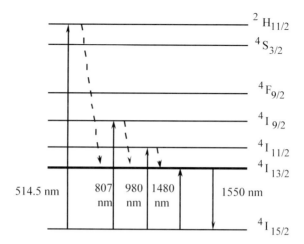

Figure 6.3. Energy level diagram of the trivalent erbium ion Er^{3+} and several pump wavelengths used for lasing in erbium doped fibre.

($^S L_J$) notation for the energy levels is used.) In radiative transitions a useful quantity is the probability of a spontaneous transition between two levels this is known as the Einstein coefficient, A_{12}, given by:

$$A_{12} = \frac{16\pi^3 n \langle v \rangle^3}{3\varepsilon_0 \, h \, c^3 g_1} K \, S_{12} \qquad (6.1)$$

where ε_0 is the permittivity of vacuum, h is Planck's constant, c the speed of light in empty space, n the refractive index of the host medium, $\langle v \rangle$ the mean photon frequency, g_1 the degeneracy of the initial state ($g_1 = 2J_1 + 1$), K a local correction, and S_{12} the line strength of a transition connecting two J multiplets.

The traditional way to characterise optical transition probabilities is the oscillator strength f_{ij}. The oscillator strength f_{12} for a transition 1→2 is a dimensionless quantity defined in terms of the Einstein coefficient A_{12}:

$$f_{12} = \varepsilon_0 m \, c^3 A_{12} / (2\pi \, n \, e^2 \langle v \rangle^2 K) \qquad (6.2)$$

where m and e are respectively the mass and charge of the electron.

When an excited state decays only by emission of photons, the observed transition rate is the sum of the probabilities for transitions to all final states. The excited state lifetime τ_1 is defined by:

$$\frac{1}{\tau_1} = \sum_j A_{1j}$$

and the branching ratio by

$$A_{12}/\left(\sum_j A_{1j}\right) = A_{12}\tau_1.$$

In the solid state, the energy levels are considerably broadened and it is often simpler to use the absorption or emission cross-section $\sigma(v)$ defined by:

$$\frac{dI}{dz} = \pm N\,\sigma(v)\,I$$

where I is the plane wave optical intensity at frequency v, z is the direction of propagation, and N is the density number of absorbing (or emitting) ions. The sign is negative for absorption. The cross section $\sigma(v)$ has the dimension of area.

The oscillator strength f_{12} is proportional to the spectral integral of the cross section σ_{12}:

$$f_{12} = \left(\frac{4\,\varepsilon_0\,m\,c\,n}{e^2\,K}\right) \times \int \sigma_{12}(v)\,dv$$

Very often it is necessary to know f_{12} only at the photon frequency where the cross section σ_M is a maximum. Introducing the effective bandwidth of the transition Δv, we have:

$$f_{12} = (4\,\varepsilon_0\,m\,c\,n/e^2\,K)\sigma_M\,\Delta v$$

The cross section can be determined from measurements of the excited state lifetime and emission spectrum.

$$\frac{1}{\tau} = A_{12} = \left(\frac{8\,\pi\,n^2}{c^2}\right) \times \int v^2\,\sigma_{12}(v)\,dv = \int I_{12}(v)\,dv \qquad (6.3)$$

$I_{12}(v)$ is the spontaneous emission rate per unit bandwidth for the $1 \to 2$ transition.

6.1.2 Example: Er^{3+} doped optical fibre amplifiers

6.1.2.1 Energy levels

The energy levels involved in the laser transition in Er^{3+} have a complex structure. The earth orbitals, labelled by the total angular momentum J are affected by the local electric field due to the components of the glass matrix and their degeneracy is removed by the Stark effect. The resulting fine structure of the energy levels depends on the electric field of the glass matrix and on the projection of the angular momentum **J** on to the field direction, giving rise to a J manifold set. The local electric fields inside the fibre core cause the $4f\ ^SL_J$ multiplets to split into a maximum of $J+1/2$ components. The number of possible sublevels depends on the symmetry of the host material. The local electric fields, which give rise to the Stark levels, vary between ionic sites and thus affect the magnitudes of the level splittings. The resultant variation of transition energies, known as *inhomogeneous broadening*, is typically of the order of 50 cm^{-1}. The transitions are also homogeneously broadened by the absorption or creation of low energy phonons. This broadening is determined by the temperature and by the material properties of the glass. In GeO_2, SiO_2 core fibre, this room temperature broadening is typically 18 cm^{-1} and for Al_2O_3, GeO_2, SiO_2 core it is 50 cm^{-1}.

The energy-level diagram of the trivalent erbium ion is shown in Figure 6.3; several pump wavelengths can be used for lasing in erbium doped fibres. The $^4I\frac{13}{2} - {}^4I\frac{15}{2}$ transition is a ground state transition at room temperature. However sufficient population inversion can be created by 807 nm pumping to the $^4I_{9/2}$ level to permit stimulated emission from the long lifetime metastable level ($\tau = 11$ ms) to the ground state for fibre amplifiers (see Figure 6.3). In the same manner, pumping at 514.5 nm or in the 950–1000 nm region or near 1.48 μm is possible.

The choice of $\lambda_p = 1.48$–1.49 μm transition corresponds to a particular pumping scheme where by the Er^{3+} ions are excited directly within the $^4I\frac{13}{2} - {}^4I\frac{15}{2}$ laser transition, which is near $\lambda_s = 1.53$ μm. Erbium-doped silica fibre provides an attractive means for optical signal amplification at $\lambda_s = 1.53$ μm with high gain and low insertion losses. The recent results indicate that multimode laser diode pump sources used for practical applications should fulfil high gain performance, making laser-diode pumped erbium fibre amplifiers highly efficient devices.

The $^4I\frac{13}{2}$ metastable state is the upper level for the transition, producing gain at 1500 nm. Any process removing Er^{3+} ions from this state, other than stimulated emission into the signal-mode, decreases the efficiency of the amplifier.

A first mechanism is the non-radiative relaxation directly to the ground state as a result of the interaction between the electrons and the dynamic lattice.

A second loss mechanism is the excited state absorption (ESA) (see Figure 6.4a). In this mechanism the ion is excited to $^4I_{9/2}$ level by a pump photon (800 nm) and relaxes non-radiatively into the $^4I\frac{13}{2}$ metastable state from which it absorbs a pump or signal photon (800 nm) to reach $^2H\frac{11}{2}$ (Zemon et al. 1991, Bastien and Sunak 1991). So a photon will be lost either by heat or by spontaneous emission at an

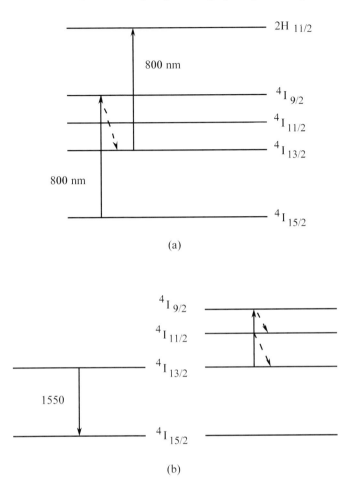

Figure 6.4. Er^{3+} doped silica fibre amplifiers (EDFAs): (a) Excited State Absorption (ESA) at 800 nm pump band, (b) Up-conversion process.

unwanted wavelength. The effect of pump-excited state absorption by Er^{3+} ions in the $^4I\,\frac{13}{2}$ excited state has been asserted to be important at some wavelengths.

ESA at 1.14 μm has also been observed when the amplifier is pumped at 528 nm (Farries 1991). No ESA occurs for 980 nm or 1480 nm pumped amplifiers.

A third cause of inefficiency is the co-operative up-conversion process (see Figure 6.5b). If two excited ions interact, one can transfer its energy to the other, itself returning to the ground state which the other is excited to the higher $^4I_{9/2}$ state. In oxide glasses the $^4I_{9/2}$ level quickly relaxes through the emission of several phonons back to the $^4I\frac{13}{2}$ level. The result is the conversion of one complete excitation into heat (Pollack and Chang 1990, Nilsson et al. 1995).

6.1.2.2 Approximate models of the levels system

Several models for EDFAs have been developed incorporating refractive index, erbium concentration profiles, and spectral distribution of amplified spontaneous emission.

The Er^{3+} doped fibre forms a system with three energy levels, or four if ESA is included (see Figure 6.5a).

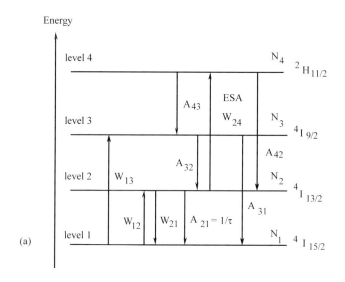

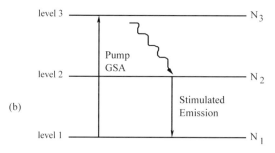

Figure 6.5. Energy levels and transitions in Er^{3+} doped fibres: (a) four-level diagram, (b) three-level diagram.

The rate equations for these populations are:

$$\begin{aligned}
\frac{dN_1}{dt} &= (W_{12} + W_{13})N_1 + (W_{21} + A_{21})N_2 + A_{31}N_3 \\
\frac{dN_2}{dt} &= W_{12}N_1 - (W_{21} + A_{21} + W_{24})N_2 + A_{32}N_3 + A_{42}N_4 \\
\frac{dN_3}{dt} &= W_{13}N_1 - (A_{32} + A_{31})N_3 + A_{43}N_4 \\
\frac{dN_4}{dt} &= W_{24}N_2 - (A_{43} + A_{42})N_4
\end{aligned} \quad (6.4)$$

The terms A_{ij} represent the total spontaneous decay rates from level i to level j, W_{1j} is the pumping rate from ground state 1, W_{24} is the ESA rate from excited level 2, and W_{12} and W_{21} are stimulated emission rates from level 1 to 2 and from level 2 to 1.

Solutions of these equations and experimental results can be found in several papers (Digonnet 1993), with four levels (Desurvire et al. 1989, Pedersen et al. 1990) with three levels (Giles and Desurvire 1991a, Ohashi and Tsubokawa 1991) or with two levels (Pedersen 1994, Giles and Desurvire 1991b, Saleh et al. 1990, Yang et al. 1989, Desurvire et al. 1989, Cognolato et al. 1994).

If we suppose $N_4 = 0$ (without ESA), three-level laser transitions can be modelled with the simplified energy level diagram of Figure 6.5b. A pump photon is observed by an electron resting in the ground state (level 1) and which is sent to the level 3 (pump band). The electron quickly relaxes to the upper laser level (level 2). From this level the electron decays down to the ground state with emission of either a spontaneous or stimulated photon. The spontaneous emission is characterised by the spontaneous lifetime τ and the stimulated emission by the effective cross section σ_e. Between levels 1 and 2, signal photon ground state absorption (GSA) also takes place, with an absorption cross section σ_a.

Since population inversion in the system is only related to the populations in the higher and lower states, it is often reasonable to simplify the three levels system to two levels.

Assuming that:

$$\tau = \frac{1}{A_{21}} > \tau_{31} = \frac{1}{A_{31}} \quad \text{and} \quad \tau_{32} = \frac{1}{A_{32}}$$

and taking $N_3 \approx 0$, $N_4 = 0$, $N_1 + N_2 = N$, using the three first equations of 6.4 gives for steady state solutions:

$$N_1 = N\frac{W_{21} + \frac{1}{\tau}}{W_{12} + W_{13} + W_{21} + \frac{1}{\tau}} \tag{6.5}$$

$$N_2 = N\frac{W_{12} + W_{13}}{W_{12} + W_{13} + W_{21} + \frac{1}{\tau}} \tag{6.6}$$

where the transition rate W_{ij} is related to the cross sections σ_a or σ_e by the relation:

$$W_{12} = \frac{\sigma_a I_s}{h\nu_s} \quad W_{21} = \frac{\sigma_e I_s}{h\nu_s} \quad W_{13} = \frac{\sigma_p I_p}{h\nu_p} \tag{6.7}$$

where σ_p is the pump absorption cross section; I_p and I_s, with frequencies respectively ν_p and ν_s, are the pump and signal photon intensities and are functions of both the radial and longitudinal position in the fibre. The reference coordinate system is a cylindrical (r, θ, z) system centred on the fibre axis with $z = 0$ at the pump input end of the fibre. For simplicity, uniform distribution of the signal in the fibre may be used:

$$I_{p,s}(r, \theta, z) = P_{p,s}(z)\frac{1}{\pi a_{p,s}^2} \quad \text{if } r \le a_{p,s} \tag{6.8}$$

where $a_{p,s}$ are effective mode radii for pump and signal.

The variations of the pump power $P_p(z)$ and forward and backward optical signal powers $P_s^+(z)$ and $P_s^-(z)$ are governed by the equations:

$$\frac{dP_p(z)}{dz} = -\sigma_p\left(\int_0^{2\pi}\int_0^\infty \frac{N_1}{\pi a_p^2}r\,dr\,d\theta\right)P_p(z) \tag{6.9}$$

$$\frac{dP_s^+(z)}{dz} = \left(\int_0^{2\pi}\int_0^\infty (\sigma_e N_2 - \sigma_a N_1)\frac{1}{\pi a_s^2}r\,dr\,d\theta\right)P_s^+(z) + P_0\sigma_e\int_0^{2\pi}\int_0^\infty \frac{N_2}{\pi a_s^2}r\,dr\,d\theta$$

$$\frac{dP_s^-(z)}{dz} = \left(\int_0^{2\pi}\int_0^\infty (-\sigma_e N_2 - \sigma_a N_1)\frac{1}{\pi a_s^2}r\,dr\,d\theta\right)P_s^-(z) - P_0\sigma_e\int_0^{2\pi}\int_0^\infty \frac{N_2}{\pi a_s^2}r\,dr\,d\theta$$

$$\tag{6.10}$$

where $P_0 = 2\,h\,\nu_s\,\Delta\nu$ is the power corresponding to spontaneous emission and the total signal power is given by $P_s(z) = P_s^+(z) + P_s^-(z)$.

By substituting equations 6.5, 6.6, 6.7 and 6.8 in 6.9 and 6.10, we obtain the following equations:

$$\frac{dP_p(z)}{dz} = -\sigma_p N \frac{a^2}{a_p^2} \left[\frac{1 + \frac{P_s}{(1+\alpha)P_{sat}}}{1 + \frac{P_s}{P_{sat}} + \frac{P_p}{P_{th}}} \right] P_p(z) \qquad (6.11)$$

$$\frac{dP_s^+(z)}{dz} = -\sigma_e N \frac{a^2}{a_p^2} \frac{\left[\left(\frac{P_p}{P_{th}} - \alpha \right) P_s^+(z) - \left(\frac{P_p}{P_{th}} + \frac{\alpha P_s}{(1+\alpha)P_{sat}} \right) P_0 \right]}{\left[1 + \frac{P_s}{P_{sat}} + \frac{P_p}{P_{th}} \right]}$$

$$\frac{dP_s^+(z)}{dz} = -\sigma_e N \frac{a^2}{a_p^2} \frac{\left[\left(\frac{P_p}{P_{th}} - \frac{2\alpha P_s}{(\alpha+1)P_{th}} \right) P_s^-(z) + \left(\frac{P_p}{P_{th}} + \frac{\alpha P_s}{(1+\alpha)P_{sat}} \right) P_0 \right]}{\left[1 + \frac{P_s}{P_{sat}} + \frac{P_p}{P_{th}} \right]} \qquad (6.12)$$

with $\alpha = \frac{\sigma_a}{\sigma_e}$, $P_{th} = \pi a_p^2 h\nu_p / \tau \sigma_p$, $P_{sat} = \pi a_s^2 h\nu_s / \tau(\sigma_e + \sigma_a)$.

In the general case, in an end pumped device configuration, the evolution of the pump power $P_p(z)$ with the fibre coordinate z follows equation 6.11 written:

$$\frac{dP_p(z)}{dz} = -\gamma_p(z) P_p(z) \qquad (6.13)$$

where $\gamma_p(z)$ is the total pump absorption coefficient including ground state (corresponding to σ_p) and excited state absorption effect (corresponding to the higher level of population density N_4).

The pump power varies according to the following equation, deduced from equations 6.11 and 6.13:

$$P_p(z) = P(0) \exp\left(-\int_0^z \gamma_p(z') \, dz' \right) \qquad (6.14)$$

6.1.2.3 Optical gain

The evolution of the output signal level relative to the input level, called the gain, varies as a function of launched pump power. In an EDFA, gain as a function of coupled pump power is measured at the gain peak wavelength. Results show that gain is saturating when the pump power increases (Choy et al. 1990, Becker et al. 1990, Zyskind et al. 1989).

Theoretical models have been presented for amplified spontaneous emission (ASE) in unsaturated EDFA (Desurvire et al. 1991). The results are useful to model ASE noise in low gain distributed-fibre amplifiers. The optical gain $g_s(z)$ and the spontaneous factor $n_{sp}(z)$ along the fibre can be obtained by the relations.

$$g_s(z) = \sigma_e N_2 - \sigma_a N_1$$

$$n_{sp} = \sigma_e N_2 / [\sigma_e N_2 - \sigma_a N_1] \tag{6.15}$$

At the output the amplifier gain G is:

$$G = G(L) = \exp\left(\int_0^L (g_s(z) - \alpha)\,dz\right) \tag{6.16}$$

where α is the absorption coefficient according to equation 6.12.

The gain coefficient is the slope of the tangent to the gain curve as a function of launched pump power, which intersects the origin. This is a widely used figure of merit to evaluate pumping efficiency. When the gain coefficient is plotted as a function of the pump wavelength for a constant pump power, it indicates a maximum value at a given signal wavelength for each fibre length (Desurvire et al. 1990). The relationship between fibre length and signal gain indicates a maximum signal gain for a particular optimum fibre length (Yamada et al. 1989).

The gain of an EDFA is a function of several factors, such as pump and signal wavelengths, fibre losses, dopants, and host glass composition. Gain and gain efficiency can be predicted and calculated in terms of these parameters (Wysocki et al. 1994, Zervas et al. 1992). Waveguide design, pump wavelength, host glass and doping concentration play a significant role in amplifier efficiency.

Temperature influences the absorption, fluorescence and small signal gain characteristics of an erbium doped fibre (Millar et al. 1990, Kagi et al. 1991). Results show that an EDFA operating near full inversion between 1550–1560 nm is stable with changing temperature over a wide operating range. However, an amplifier operating at peak gain near full inversion is sensitive to variations in ambient temperature.

Polishing the fibre input end at an angle may reduce the competition between the lasing at the gain peak and the amplified signal of the gain peak. Saito et al. 1991 have demonstrated the importance of reflections for high gain operation.

The dopant concentration also has an effect on the optical amplification characteristics (Shimizu et al. 1990). In several applications, such as wavelength division multiplexing, broadening the gain bandwidth is necessary. Wider bandwidths are achieved by adding Al_2O_3 or P_2O_5 to silica glass as co-doping materials (Yamada et al. 1990a) or by using two kinds of fibres in line amplifiers, such as silica glass and ZBLAN (zirconium barium lanthanum) glass amplifiers.

6.1.2.4 Noise

As mentioned earlier, there is a growing interest in the use of EDFAs at 1.5 μm for optical communications. For applications with a large number of amplifiers it is necessary to know the degradation of the signal precisely. Several papers are devoted to modelling or measuring the noise (Kikuchi 1993, Martin and Duan 1994, Gianfrango et al. 1993, Giles and Desurvire 1991a, Yamada et al. 1990b, Laming and Payne 1990, Chen and Desurvire 1992, Clesca et al. 1994a, Masuda et al. 1993, Bonnedal 1993, Nykolak et al. 1991).

Like gain, noise also varies with wavelength and input power level. These characteristics are not trivial; when a source covers a wavelength range of 1.5–1.6 μm, the output from a high-gain amplifier may cover a dynamic range of 30–40 dB. However, the amplified spontaneous emission (ASE) is always present. Since spontaneous emission is present together with stimulated processes, optical amplification is unavoidably combined with generations of spontaneously emitted photons within the fluorescence band. The uncorrelated photon flow partially couples into the guided mode and is superimposed on the coherent beam of the amplified signal, constituting a noise source. Therefore, a laser travelling wave optical amplifier necessarily degrades the signal-to-noise ratio (S/N).

The performance of an amplifier can be indicated by the noise figure. It is the ratio of input to output S/N ratios:

$$F = (S/N)_{\text{in}}/(S/N)_{\text{out}} \quad \text{or} \quad NF = 10 \ \log_{10}(F) \qquad (6.17)$$

The noise figure of an optical amplifier is made up of four contributions, namely the shot noise of the amplified signal and the ASE, and the beat noise due to heterodyning of signal and broadband ASE, and of broadband ASE with itself. Considering the simplified diagram (Figure 6.5b) showing the energy levels of erbium (three-level scheme), optical pumping is applied at 980 nm (corresponding to energy between levels 1 and 3). The equations governing the population variation are equations 6.4. Equations 6.5 and 6.6 give the populations at the two levels 2 and 1.

The spontaneous emission factor $n_{\text{sp}}(z)$ and the optical gain $g_{\text{s}}(z)$ along the fibre can be calculated from equation 6.15.

At the output the noise power spectral density N_{P} is given by:

$$N_{\text{P}} = N_{\text{P}}(\omega_0) = G(L)\hbar \ \omega_{\text{s}} \int_0^L [g_{\text{s}}(z) n_{\text{sp}}(z)/G(z)] \ dz \qquad (6.18)$$

where $G(z)$ is given by equation 6.16, and the symbol $\hbar$ stands for $h2\pi$ (where h is Planck's constant).

The output noise power is obtained by adding the shot noise due to the amplified signal and due to spontaneous emission:

$$\sigma^2 = \left[2G\langle P_{\text{in}}\rangle N_P + 4N_P^2 \Delta f + G\langle P_{\text{in}}\rangle \hbar\,\omega + 2N_P \Delta f\,\hbar\,\omega\right]2B_e$$

where $\langle P_{\text{in}}\rangle$ is the power spectral density at the input, Δf the spontaneous emission bandwidth, and B_e the bandwidth of the electrical signal. From the measured gain and ASE spectra, the noise figure may be estimated. It is calculated as:

$$F = \frac{2N_P + 1}{G} = \frac{2N_P}{G} + \frac{1}{G} \tag{6.19}$$

where the dimensionless noise term N_P can also be written as:

$$N_P = \frac{\lambda^3}{2hc^2}ASE$$

Here, ASE is the unpolarised spectral power per wavelength of the spontaneous emission.

Noise due to backward travelling ASE can be suppressed by an isolator (Lumholt *et al.* 1992, Laming *et al.* 1992a).

6.1.3 Other doped optical fibres

Several other rare-earth doped fibres have been studied. To date, most of the attention has been focused on erbium, neodymium and praseodymeum doped fibres (Furthner and Penzkofer 1992) primarily because of their promise as direct optical amplifiers in the communications bands near 1.5 µm and 1.3 µm respectively. We have in § 6.1.2 described the EDFA as an example, and we showed (Figure 6.2) the energy level diagram for neodymium. The fluorescence bands of Nd^{3+} at 920, 1070 and 1340 nm correspond to the $^4I_{9/2}, ^4I\,\frac{11}{2}$ and $^4I\,\frac{13}{2}$ levels respectively. The strongest band is centred at 1100 nm and has a 3 dB bandwidth of 55 nm. However, five other rare-earths have successfully been incorporated into silica hosts and operated as fibre laser. These may be used in other application areas, such as sensors or sources with specific wavelengths. In recent years theoretical and experimental work has been carried on four rare-earths: ytterbium Yb^{3+} (Magne *et al.* 1994a, Ouerdane *et al.* 1994); holmium (Ho^{3+}) (Saissy *et al.* 1991); praseodymium (Pr^{3+}) (Whitley 1995); and thulium (Tm^{3+}) (Gomez *et al.* 1993). Other rare-earths for doped fibres are also promising (Magne *et al.* 1994b).

An alternative to silica glass is the fluoride glass known as ZBLAN, using the fluorides of zirconium (Zt), barium (Ba), lanthanum (La), aluminium (Al) and sodium (Na), doped with rare-earth ions. Such glasses exhibit a broader fluorescence spectrum than conventional silica glass when used in optical fibres. Consequently, such a fibre can amplify signals over a broader region than a silica fibre. This can prove an advantage in wavelength division multiplexing (WDM) amplifying systems. For example, erbium-doped ZBLAN fibre amplifiers display a flatter amplified spontaneous emission (ASE) than conventional EDFAs, with a bandwidth around 25 nm and gain, S/N ratio and pump efficiency close to that of silica glass. They can be used for WDM-tunable emission. Continuous-wave laser emission has been reported in erbium doped fluorozirconate fibre (Brierley and France 1988, Auzel et al. 1988, Allain et al. 1989a, Allain et al. 1989b, Tesar et al. 1992). In Nd^{3+} the $^4F_{3/2} \rightarrow \, ^4I\frac{13}{2}$ transition (around 1340 nm) suffers from signal excited state absorption (ESA), which limits the gain to wavelengths longer than 1,36 μm. The use of fluoride fibre as host reduces significantly this problem (Ammann et al. 1994). Investigative studies compare the efficiency of fluoride glasses with a variety of compositions (Zemon et al. 1992, Rasmussen et al. 1992).

The Pr^{3+} doped fluoride fibre is promising, owing to its ability to produce a broad-band light sources with high gain in 1,3 μm telecommunications systems (Ohishi et al. 1991, Ohishi et al. 1992, Sugawa and Miyajima 1991). Gain and noise in Pr^{3+} doped fluoride fibre have also been investigated for variables parameters such as core diameter, numerical aperture, input pump power and dopant concentrations (Pederson et al. 1992, Karasek 1993, Karasek 1994).

6.1.4 Device aspects of fibre amplifiers, performance and applications

The three ways of using a fibre amplifier in telecommunications systems are depicted in Figure 6.6. The signal is launched into one end of the doped fibre and collected from the other end.

In a booster amplifier (see Figure 6.6a). The signal is amplified at the bottom of the telecommunication system. One application could be the simultaneous amplification of many channels in a multichannel LAN (Local Area Network) transmitter station. A high saturation output power would be required.

Repeater (see Figure 6.6b). In systems where chromatic dispersion is negligible (coherent system or LAN), amplifiers could favourably replace optoelectronic repeaters which have a much smaller bandwidth.

Pre-amplifier (see Figure 6.6c). Used with a high speed PIN-FET receiver, pre-amplifier will allow the operation of a direct detection system at data rates above 5 Gbit/s.

Optical amplification technology plays an important role in 1.5 μm wavelength optical communication systems. In practical applications, laser diode pumping techniques are indispensable. Figure 6.7 shows a schematic diagram of a simple EDFA module. It consists of four optical components: a length of erbium doped fibre, a pumping LD module, a wavelength-division multi/demultiplexing (WDM) fibre coupler and a polarization-insensitive-type optical isolator with fibre pigtail. This isolator is inserted within the active fibre in order to reduce the accumulation of backward travelling amplified spontaneous emission (ASE). At the signal input end ASE, as a low noise figure depends upon

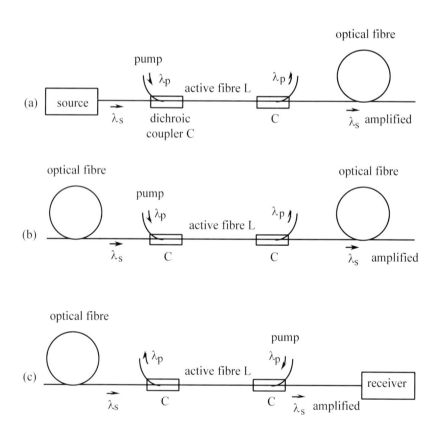

Figure 6.6. Three types of fibre amplifier: (a) power amplifier (booster), (b) repeater, (c) pre-amplifier.

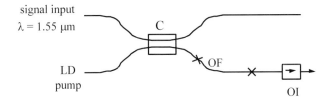

C	0.98/1.55 µm coupler or 1.48/1.55 µm coupler,	OI	Optical isolator,
OF	Erbium doped fibre, LD	Laser diode pump. $\lambda = 0.98$ µm or $\lambda = 1.48$ µm	

Figure 6.7. Schematic of an EDFA.

a high population inversion in this part of the fibre (Lumholt et al. 1992). Gains in the range 26–40 dB are obtained with 1.48 µm pumping and 26–34 dB with 0,98 µm pumping. When the EDFA is placed at the end of the transmission line before the receiver, a very high gain (54 dB) and a noise figure close to the quantum limit of 3.1 dB are obtained (Laming et al. 1992a). A two-stage pre-amplifier with a sensitivity of 38.8 dBm at 10 GB/s has been demonstrated (Laming et al. 1992b, Shimizu et al. 1990).

An isolator can also be used within the erbium doped fibre to avoid the EDFA from acting as a laser, by suppression of external reflections (Lester et al. 1995).

Several experiments that combine WDM with optical amplification in either long distance or distribution systems, have shown good system performance. The main difficulty is to keep the gains and noise level independent of wavelength. Several methods have been used for equalising the gain of the silica fibre amplifier: co-doping of the host matrix with aluminium and lanthanium; optical filtering; spatial hole burning; use of a ring laser and use of a hybrid amplifier composed of an erbium doped silica fibre joined to an erbium doped ZBLAN fibre (Clesca et al. 1994b, Willner and Hwang 1995, Tachibana et al. 1991, Forghieri et al. 1995, Eskildsen et al. 1993, Bayart et al. 1994, Semenkoff et al. 1994, Ali et al. 1994, Inouee et al. 1991). Some works have proposed the use of a twin-core erbium doped optical fibre amplifier as a passive channel equaliser (Zervas and Laming 1995) or, alternatively, adjusting the pump power into one of the core in order to tune the gain characteristic (Wu and Chu 1994).

Amplification of ultra-short solitons in EDFA is possible for data transmission over several hundred kilometres (Olsson et al. 1990). Models including both gain saturation and gain dispersion have also been studied (Agrawal 1990, Chi et al. 1994).

6.2 RARE-EARTH DOPED FIBRE LASERS (REDFLs)

In addition to their applications as optical amplifiers, rare-earth doped fibres have been used in optical resonators to create fibre lasers (REDFLs), and, with wavelength-selective elements, to achieve tunable laser operation over an amplified broad fluorescence band.

6.2.1 Principle

Bearing in mind the high powers necessary to achieve non-linear responses, it is important to note that the damage power limit for pure silica is about 10^{10} W cm^{-2}, which corresponds to a power of 5 kW for an 8 μm core diameter.

Investigations of lasing action in which a fibre is the active medium have used Fabry-Pérot resonators or fibre loop reflectors. A fibre Fabry-Pérot laser is longitudinally pumped by an external source, which is usually some readily available line from a commercial laser. Ideally a semiconductor diode laser should be used as a pump, as it is compact and has a high efficiency.

A simple experimental set-up of a REDFA using mirrors is shown in Figure 6.8. A pump source is launched into a fibre laser through a mirror M_1. This mirror M_1 should be highly transparent at the pump wavelength and highly reflective at the signal wavelength. A dichroic mirror M_2 with almost 100% reflectance at the pump laser wavelength and 95% transmittance at 1.5 μm is suitable (Iwatsuki 1990). However, M_2 can also be butted against the fibre end face (Schneider 1995) or a reflection grating can be used instead of mirror M_2 (§ 6.2.2). After the mirror M_2 a dichroic beamsplitter or an ordinary beamsplitter and a bandpass filter are inserted, transmitting the laser wavelength λ but reflecting any remaining pump energy out of the system (or simply absorbing it). The fibre must have a single transverse mode at both pump and signal wavelengths to maximize the power confinement and the overlap of the pump and signal beams.

Optical excitation is initiated by focusing light into a fibre end through a mirror consisting of dielectric thin film coatings on a substrate. A difficulty is the loss caused by the end alignment. When using a high numerical aperture objective for launching, it is necessary for the lens front surface to be close to the fibre end, to maximize the launching efficiency. Therefore the mirror substrate must be thin. A second difficulty is possible thermal damage to the mirror coatings when the fibre is pumped with a high power laser. The output light from a fibre Fabry-Pérot laser is usually continuous wave (CW) and has a bandwidth typically of the order of 1 to 10 nm.

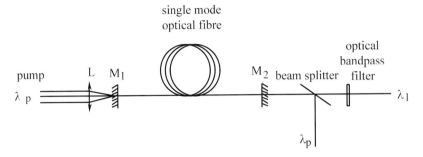

Figure 6.8. Schematic of a fibre laser: λ_p pump wavelength, λ_1 fibre laser wavelength, M_1, M_2, mirrors with coatings and substrate, L microscope objective.

Erbium doped fibre lasers (EDFLs) can operate spontaneously (self-pulsing) as well in CW a (Le Boudec *et al.* 1994).

6.2.2 Fibre laser using gratings

An experimental set-up of an EDFL using gratings is shown on Figure 6.9a. A pump source is launched into an erbium-doped fibre through a mirror. After the end of the fibre, a grating, which provides feedback and significantly reduces the linewidth of the laser output is incorporated (Jauncey *et al.* 1987). A thin layer of index-matching fluid is placed on this grating in order to increase the interaction with the evanescent field of the fibre.

Much effort has been made to narrow the linewidths of erbium or neodymium doped fibre lasers by using a ring cavity with polarization-holding fibres (linewidth ≈ 1.4 KHz) (Iwatsuki *et al.* 1990) fibre gratings in the Fabry-Pérot cavity or Fox-Smith resonators (see Figure 6.10b) (linewidth ≈ 1 MHz). In the first case, an optical bandpass filter is incorporated into the ring to make the lasing wavelength tunable. A Fox-Smith resonator consists of two cavities of different lengths with a common part between the coupler and one mirror. The two cavities are coupled and there is an interchange of power between them. The resulting longitudinal mode structure depends on the lengths of both cavities which are not exactly equal. The cleaved fibre ends are directly butted to dielectric mirrors having a reflectance greater than 99.5% at the lasing wavelength and a transmittance greater than 80% at the pump, wavelength. Good results can be obtained by using Bragg gratings written in to the fibres (Digonnet 1993).

Permanent holographic gratings have been written in many commercial optical fibre types with germanium-doped cores. These

240 *Lasers and amplifiers based on rare-earth doped fibres*

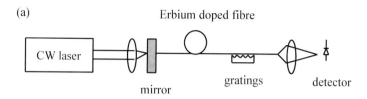

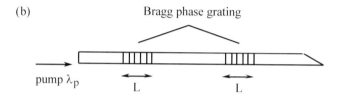

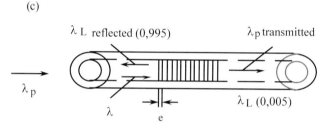

Figure 6.9. Doped fibre laser using gratings: (a) external gratings, (b) Bragg phase gratings written holographically into the core $L = 1$ or 2 cm, (c) Illustration showing fibre grating reflector and the Bragg condition for reflection.

include the standard telecommunications fibres, and the rare-earth doped amplifier or laser fibres. The gratings are made with a technique using interfering UV laser beams on the side of the fibre. Several devices are made with such gratings: filters, reflectors, shifters, multiplexers, etc. (see Chapter 3). One such device is an embedded grating. Two Bragg phase gratings written holographically into the core are used (Figure 6.9b). This is achieved by removing the sheath from a part of the fibre and exposing each part transversely to a two UV beam interference pattern. The gratings are written at 10 cm from each cleaved end of a 0.5 m fibre (Ball *et al.* 1991). An Er^{3+} doped fibre main cavity and auxiliary silica fibre cavity made with three intra-core Bragg gratings can be useful for operation as a self mode-locked laser (Cheo *et al.* 1995).

Cavities of high birefringence REDFLs have been closed by an intracore Bragg grating reflector and a dichroic mirror (Pureur *et al.*

1995, Douay et al. 1992). These gratings are used as reflectors following the Bragg condition $\lambda_L = 2\Lambda_B$, where Λ_B is the grating spacing (Figure 6.9c)

6.2.3 Fibre laser using directional couplers

Several devices can be constructed from directional couplers, fibres and mirrors (see Figure 6.10). In a transversely coupled fibre Fabry-Pérot

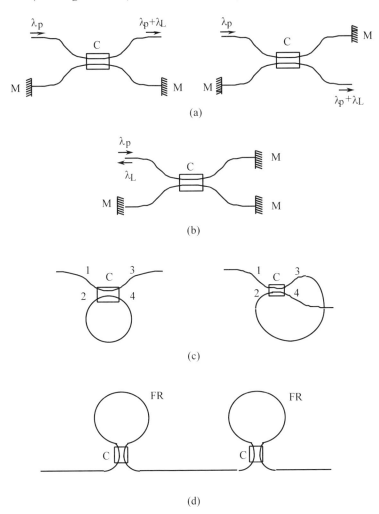

Figure 6.10. Fibre laser using directional couplers, C coupler, M mirror, FR fibre ring: (a) transversely coupled fibre resonator, (b) Fox-Smith resonator, (c) using fibre ring resonator, (d) using two fibre ring reflectors.

resonator, the pump light is launched into a fibre end that does not have mirrors. It is the same for lasing light (Figure 6.10a). The Fox–Smith resonator is applied to longitudinal mode selection of the lasing response (see Figure 6.10b). Three of the fibre ends have mirrors. Butting fibre ends against bulk mirrors results in losses. It would be preferable to employ a fibre laser resonator without a mirror, allowing the independent choice of an output coupling condition from the two fibre ends, two ports of a fibre directional coupler are spliced together. In Figure 6.10c. The splice is avoidable by folding the fibre in the appropriate manner before the coupler is made. A fibre-based interferometer that overcomes the problem, is an all-fibre loop reflector. Two such loop reflectors are joined in series (see Figure 6.10d).

The fibre loop reflector is a non-resonant device. Light launched into one of the ports of the loop is split through the directional coupler, circulates both clockwise and counterclockwise round the loop and then coherently recombines at the coupler (see Chapter 3 and Figure 3.39). The output wave must exit by either the input port, in which case it is a reflected wave, or by the opposite port in which case it is a transmitted wave. The two light paths have the same length and there is no phase shift associated with unequal propagation distances. The reflectance R, transmittance T and loss A are given by:

$$R = 4CR(1 - CR)(1 - \alpha_c)^2 \exp(-2\alpha L)$$
$$T = (1 - 2CR)^2(1 - \alpha_c)^2 \exp(-2\alpha L) \quad (6.20)$$
$$A = 1 - (1 - \alpha_c)^2 \exp(-2\alpha L)$$

where L is the loop length, α_c and α the coupler loss and fibre loss, and CR is the intensity coupling ratio of the coupler.

The fabrication of many ring single-mode fibre laser devices has been achieved using low-loss erbium-doped fibres (Zhang and Lit 1994, Shi *et al.* 1995, Okamura and Iwatsuki 1991, Pfeiffer *et al.* 1992). Industrial devices are already available.

A fibre-ring laser includes a coupler and an isolator. The configuration gives a stable setting of the lasing wavelength and a single-mode operation with less than 10 kHz linewidth (Cowle *et al.* 1991).

6.2.4 Fibre laser using fibre reflectors

A fibre loop can be a component of a fibre laser (see Figure 6.11). The resonant optical cavity uses two fibre reflectors, each formed by a single fibre loop between the output ports of a directional coupler (see

Figure 6.10d). The mirrors are fabricated from a continuous length of doped fibre formed into a coupler loop. Light that couples across such a coupler undergoes a π/2 phase lag. The loss A, the intensity transmittance T and reflectance R in the coupler loop are given by equation 6.20.

When two couplers are joined, they form an all-fibre resonator. The intensity response is:

$$\frac{I_0}{I_1} = \frac{T_1 T_2 \exp(-2\alpha L_3)}{R^2 \left[1 + (4R_1 R_2/R^2) \sin^2 B\right]} \quad (6.21)$$

with $R = 1 - R_1 R_2$
$B = \beta[(L_1 + L_2)/2 + L_3]$
β is the propagation constant of the mode.

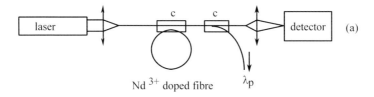

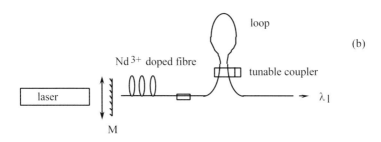

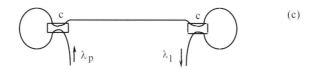

Figure 6.11. Several arrangements for a REDFL using fibre loops: (a) with loop and couplers, (b) with mirror and fibre loop reflector, (c) with two fibre loop reflectors.

Where I_0 is the output intensity, I_i the input intensity, R_1, T_1, L_1 the parameters of the first coupler, R_2, T_2, L_2 the parameters of the second coupler and L_3 and α respectively the length and the loss of the fibre without the two couplers. In the cavity described in Figure 6.11d there is a $\pi/2$ phase change on each reflection. An example using a Nd^{3+} fibre laser is given by Miller and et al. 1987.

6.2.5 Q-switching fibre laser

There are many applications in which it is desirable to achieve narrow line operation with the option of wavelength tunability, or to be able to have the radiation in the form of pulses with high peak powers or short durations, which are provided by Q-switching and mode locking (Schlager et al. 1989).

It is often necessary to incorporate a beamsplitting element so that the cavity can be accessed by means other than through the end reflectors. In optical fibre resonators, beamsplitting is carried out using directional couplers. A method of modifying a fibre Fabry-Pérot resonator is to couple light out of the fibre using a lens or microscope objective at some point between the two mirrors. Elements to deflect, modulate, polarize, reflect or diffract light at the lasing wavelength may be inserted into the beam (see Figure 6.12). These devices have

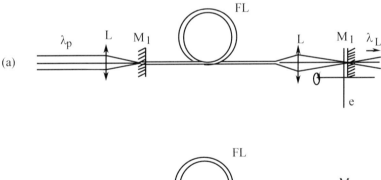

Figure 6.12. Methods for Q-switching: (a) mechanical, (b) acousto-optic, M_1, M_2 mirrors, L microscope objective, e chopper, AOD acousto-optic device, λp, λ_L pump and laser wavelength, FL fibre laser.

been successfully used to Q-switch, mode lock and tune fibre lasers (Barnes et al. 1989).

The usual methods of Q-switching lasers, namely mechanical, acousto-optic devices (AODs), electro-optic and absorption devices, may be applied to fibre lasers. Among these the first two are the most common.

Lasers may be Q-switched with a mechanical chopper (see Figure 6.12a). It provides a very high extinction ratio and has zero insertion loss in the high Q state. However, it has a slow switching time and poor pulse-to-pulse stability. By using an acousto-optic deflector a faster modulation of the cavity can be achieved (see Figure 6.12b). Careful choice of the acousto-optic material, together with anti-reflection coating, may reduce the insertion loss of the modulator, so low-threshold operation may be possible. However, a difficulty associated with AODs is the limited diffraction efficiency. It may be used in the zero order configuration or in the first order configuration. In the first order configuration, feedback is obtained by positioning the output coupler to reflect the first order diffracted beam back to the fibre. When the AOD is 'off' there is no deflection and thus no CW lasing.

6.2.6 Mode-locking fibre laser

The aim of mode locking a fibre laser is to create pulses with shorter durations than those obtained from Q-switching or gain switching. Mode locking can be achieved by either passive or active means. In passive mode locking, a saturable absorbing element is inserted into the cavity. The beam whithin the doped optical fibre is absorbed until the intensity threshold is reached for the element. The transmission time (less than 1 ps) is shorter than the oscillation time of the cavity. The saturable absorbing element is driven directly by the optical beam.

In active mode locking, an external driving signal is applied to a modulator. This element is either an electro-optic material ($LiNbO_3$) or an acousto-optic material (quartz). Two kinds of active mode locking can be used in a cavity: Amplitude (AM) or Frequency (FM) Modulation by a phase modulator. The principle of a phase modulator was described in Chapter 3 for standard fibres. It also applies to REDFLs.

6.2.7 Tunable operation

The broad emission lines of impurity ions in glass offer the possibility of a wide tuning range, suggesting that fibre lasers may prove to be a new class of tunable lasers. They have the advantages of good photochemical stability and room temperature operation. A fibre laser is tuned by one of several possible techniques.

The first of these consists in changing the spectral dependence of the intensity coupling ratio of a fused fibre coupler in a loop reflector by the thermo-optic effect. A shift in the reflectance of the cavity output coupler of a fibre laser as a function of wavelength is induced. The fibre loop reflector is spliced to the fibre laser and the temperature of the coupler changes the reflectance, which is function of the coupling ratio CR. The temperature change perturbs the interacting mode fields in the coupler; the reflectance is a function of wavelength and temperature (Millar et al. 1988). An example of this structure is shown in Figure 6.13 with two coupled cavities (Fox-Smith resonator).

A second possibility consists of the installation inside the resonator, between the fibre end and the output coupler, of an objective, two quarter-wave plates and a birefringent coupler (Alcoch et al. 1986). Tuning a fibre laser can also be achieved by using a grating as the frequency selector between the fibre and the output coupler.

The maximum value $R_M(\lambda)$ of the reflectance coefficient $R(\lambda)$ is:

$$R_M(\lambda) = (1 - \alpha_c)^2 \exp(-2\alpha \, L) \tag{6.22}$$

corresponding to CR = 1/2 in equation 6.20. Hence the spectral position of maximum reflectance is tuned by changing CR. To construct the tunable reflector, the length of the overcoupled fused fibre region is embedded into a low refractive index silicone compound. The index of this compound varies with temperature.

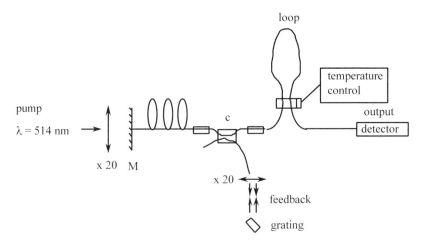

Figure 6.13. Fox-Smith resonator arrangement [After Millar et al. 1988].

Another method is based on an intracavity polarizer inserted into a Fabry-Pérot doped fibre laser (Ghera et al. 1992). The tunability is achieved by either rotating the polarizer, or varying the fibre birefringence. Wavelength selection in the fibre ring laser can be realized by an element such as a liquid crystal or angle-tuned etalon filter (Maeda et al. 1990, Pfeiffer and Veith 1994), or an acousto-optic tunable filter (Frankel et al. 1994, Cognolato et al. 1995).

Use of two sections of different Er^{3+} doped fibres can also enable wavelength selection and tuning over a range of 8 nm (Loh et al. 1994).

6.2.8 Superfluorescent rare-earth doped fibre sources

In the previous section, wavelength selection has been demonstrated by means of different tunability operations. For superfluorescent fibre source, wavelength selection does not occur, but for each wavelength in the fluorescence band of the rare-earth ion, power is amplified. This is because when pump power is inserted in a length of REDF spontaneous emission occurs in the emission band. If the level of the pump power is high enough, spontaneous amplification leads to stimulated emission. If the pump power is sufficiently high there will be no need for mirrors and no resonance occurs, so these sources can be with a large spectral bandwidth.

REFERENCES FOR CHAPTER 6

Agrawal GP 1990 *IEEE Photonics Technol. Lett.* **2** 875–7
Ainslie B J 1991 *J. Lightwave Technol.* **9** 220–7
Alcock IP, Ferguson AI, Hanna DC and Tropper AC 1986 *Opt. Lett.* **11** 709–11
Ali MA, Elrefaie AF, Wagner RE, Mendez F, Pan J and Ahmed SA 1994 *IEEE Photonics Technol. Lett.* **6** 1039–42
Allain JY, Monerie M and Poignant H 1989a *Electron. Lett.* **25** 28–9
Allain JY, Monerie M and Poignant H 1989b *Electron. Lett.* **25** 318–19
Ammann H, Hodel W and Weber HP 1994 *Optics Commun.* **113** 39–45
Artiglia M, Di Vita P and Potenza M 1994 *Opt. Quant. Elect.* **26** 585–608
Auzel F, Meichenin D and Poignant H 1988 *Electron. Lett.* **24** 909–10
Ball GA, Morey WW and Glenn WH 1991 *IEEE Photonics Technol. Lett.* **3** 613–15
Barnes B, Morkel P, Farries M, Reekie L and Payne D 1989 *Fibre Laser Sources and Amplifiers SPIE* **1171** 298–308
Bastien SP and Sunak HRD 1991 *IEEE Photon. Technol. Lett.* **3** 456–8

Bayart D, Clesca B, Hamon L and Beylat JL 1994 *IEEE Photonics Technol. Lett.* **6** 613–15

Becker PC, Lidgard A, Simpson JR and Olsson NA 1990 *IEEE Photonics Technol. Lett.* **2** 35–7

Bonnedal D, 1993 *IEEE Photonics Technol. Letters* **5** 1193–6

Brierley MC and France PW 1988 *Electron. Lett.* **24** 935–7

Chen DN and Desurvire E 1992 *IEEE Photon. Technol. Lett.* **4** 52–5

Cheo PK, Mutalik VG and Ball GA 1995 *IEEE Photonics Technol. Lett.* **7** 980–2

Chi S, Chang CW and Wen S 1994 *Optics Commun.* **106** 193–6

Choy MM, Chen CY, Andrejco M, Saifi M and Lin C 1990 *IEEE Photonics Technol. Lett.* **2** 38–40

Clesca B, Bousselet P, Augé J, Blondel JP and Février H 1994a *IEEE Photonics Technol. Lett.* **6** 1318–20

Clesca B, Ronarch'h D, Bayart D, Sorel Y, Hamon L, Guibert M, Beylat JL, Kerdiles JF and Semenkoff M, 1994b *IEEE Photonics Technol. Lett.* **6** 509–11

Cognolato L, Gnazzo A, Sordo B and Brushi C 1994 *J. of Opt. Commun.* **15** 150–4

Cognolato L, Gnazzo A, Sordo B and Brushi C 1995 *J. of Opt. Commun.* **16** 122–5

Cowle GJ, Payne DN and Reid D 1991 *Electron. Lett.* **27** 229–30

Desurvire E, Giles CR and Simpson JR 1989 *J. Lightwave technol.* **7** 2095–104

Desurvire E, Sulhoff J W, Zysking J L and Simpson J R 1990 *IEEE Photonics Technol. Lett.* **2** 653–5

Desurvire E, Zirngibl M, Presby H M and Digiovanni D 1991 *IEEE Photonics Technol. Lett.* **3** 127–9

Digonnet JF, 1993 (Marcel Dekker: New York)

Douay M, Feng T, Bernage P, Niay P, Delevaque E and Georges T 1992 *IEEE Photonics Techn. Lett.* **4** 844–6

Eskildsen L, Goldstein E, Da Silva V, Andrejco M and Silberberg Y 1993 *IEEE Photonics Technol. Lett.* **5** 1188–90

Farries MC 1991 *IEEE Photonics Technol. Lett.* **3** 619–20

Forghieri F, Tkach RW and Chraplyvy AR 1995 *J. of Lightwave Technol.* **13** 889–97

Frankel MY, Esman RD and Weller JF 1994 *IEEE Photonics Technol. Lett.* **6** 591–3

Furthner J and Penzkofer A 1992 *Opt. and Quantum Electronics* **24** 591–601

Ghera U, Konforti N and Tur M 1992 *IEEE Photon. Technol. Lett.* **4** 4–6

Gianfrango L, Cariolaro, Corvaja R, Franco P, Michio M and Pieroben G 1993 *Fibre and Integrated Opt* **13** 199–213

Giles CR and Desurvire E 1991a *J. Lightwave Technol.* **9** 147–54
Giles CR and Desurvire E 1991b *J. Lightwave Technol.* **9** 271–83
Gomez ASL, Boyer GR, Demouchy G, Mysyrowic Z, Poignant H and Monerie M 1993 *Optics Commun.* **95** 246–50
Inoue K, Kominato T and Toba H 1991 *IEEE Photonics Technol. Lett.* **3** 718–20
Iwatsuki K 1990 *IEEE Photonics Technol. Lett.* **2** 237–8
Iwatsuki K, Okamura H and Saruwatari M 1990 *Electron. Lett.* **26** 2033–4
Jauncey IM, Reekie L, Mears RJ and Rowe CJ 1987 *Opt. Lett.* **12** 164–5
Kagi N, Oyobe A and Nakamura K 1991 *J. Lightwave Technol.* **9** 261–5
Karasek M 1993 *Optics Commun.* **96** 55–8
Karasek M 1994 *Optics Commun.* **107** 235–9
Kikuchi K 1993 *Fibre and Integrated Optics* **12** 369–80
Laming RI, Barnes WL, Reekie L, Morkel PR, Payne DN and Vodhanel RS 1989 *SPIE* **1171** 82–92
Laming RI and Payne DN 1990 *IEEE Photonics Technol. Lett.* **2** 418–21
Laming RI, Zervas MN and Payne DN 1992a *IEEE Photonics Technol. Lett.* **4** 1345–7
Laming RI, Gnauck AH, Giles CR, Zervas MN and Payne DN 1992b *IEEE Photonics Technol. Lett.* **4** 1348–50
Le Boudec P, Sanchez F, Francois PL, Bayon JF and Stéfan GM 1994 *Annales des Télécommunications* **49** 178–92
Lester C, Schüsler K, Pederson B, Lumholt O, Bjarklev A and Povlsen JH 1995 *IEEE Photonics Technol. Lett.* **7** 293–5
Loh WH, Morkel PR and Payne DN 1994 *IEEE Photonics Technol. Letters* **6** 43–6
Lumholt O, Schüsler K, Bjarklev A, Dahl-Peterson S, Povlsen JH, Rasmussen T and Rottwitt K 1992 *IEEE Photonics Technol. Lett.* **4** 568–70
Maeda MW, Patel JS, Smith DA, Lin C, Saifi MA and Von Lehman A 1990 *IEEE Photonics Technol. Lett.* **2** 787–9
Magne S, Druetta M, Goure JP, Thévenin JC, Ferdinand P and Monnom G 1994a *J. of Luminescence* 60–1
Magne S, Boisdé G, Monnom G, Ouerdane Y, Druetta M, Goure JP and Mazé G 1994b *Journal of Physique* IIC **4** 451
Martin S, Duan GH, 1994 *Ann. Telecommun.* **49** 490–8
Masuda H, Aida K and Nakagawa K 1993 *IEEE Photonics Technol. Lett.* **5** 1436–8
Millar CA, Miller ID, Mortimore DB, Ainslie BJ and Urquhart P 1988 *IEE Proc PtJ.* **135** 303–9
Millar CA, Whitley TJ and Fleming SC 1990 *IEE Proc. PtJ* **137** 155–62

Miller ID, Mortimore DB, Urquhart P, Ainslie BJ, Craig SP, Millar CA and Payne DB 1987 *Appl. Opt.* **26** 2197–201

Nilsson J, Blixt P, Jaskorzynska B and Babonas J 1995 *J. Lightwave Technol.* **13** 341–9

Nykolak G, Kramer SA, Simpson JR, DiGiovanni DJ, Giles CR and Presby HM 1991 *IEEE Transactions Photonics Technol. Lett.* **3** 1079–81

Ohashi M and Tsubokawa M 1991 *IEEE Photonics Technol. Lett* **3** 121–3

Ohishi Y, Kanamori T, Nishi T and Takahashi S 1991 *IEEE Photonics Technol. Lett.* **3** 715–17

Ohishi Y, Kanamori T, Nishi T, Takahashi S and Snitzer E 1992 *IEEE Photonics Technol. Lett.* **4** 1338–41

Okamura H and Iwatsuki K 1991 *J. of Lightwave Technol.* **9** 1554–60

Olsson NA, Andrekson PA, Becker PC, Simpson JR, Tanbun-ek T, Logan RA, Presby H and Wecht K 1990 *IEEE Photonics Technol. Lett.* **2** 358–9

Ouerdane Y, Magne S, Druetta M, Boukenter A, Gourc JP and Jacquier B 1994 *J. de Physique IVC* **4** 545–8.

Payne DN 1992 *Fibre and Integrated Optics* **11** 191–219

Pedersen B 1994 *Opt. Quant. Elect.* **26** 273–84

Pedersen B, Miniscalco WJ and Quimby RS 1992 *IEEE Photonics Technol. Lett.* **4** 446–8

Pedersen B, Dybdal K, Dam Hansen C, Bjarklev A, Povlsen JH, Vendeltorp-Pommer H and Larsen CC 1990 *IEEE Photonics Technol. Lett.* **2** 863–5

Pfeiffer Th, Schmuck H and Bülow H 1992 *IEEE Photonics Technol. Lett.* **4** 847–9

Pfeiffer Th and Veith G 1994 *Optical and Quantum Electron.* **26** 547–57

Pollack SA and Chang DB 1990 *Optical and Quantum Elect.* **22** 75–93

Pureur D, Douay M, Bernage P, Niay P and Bayon JF 1995 *J. of Lightwave Technol.* **13** 350–5

Rasmussen T, Bjarklev A, Lumholt O, Obro M, Pederson B, Povlsen JH and Rottwitt K 1992 *IEEE Photonics Technol. Lett.* **4** 49–51

Saissy A, Ostrowsky DB and Maze G 1991 *J. of Lightwave Technol.* **9** 1467–70

Saito T, Sunohara Y, Fukagai K, Ishikawa S, Henmi N, Fujita S and Aoki Y 1991 *IEEE Photonics Technol. Lett.* **3** 551–3

Saleh AAM, Jopson RM, Evankow JD and Aspell J 1990 *IEEE Photonics Technol. Lett.* **2** 714–17

Schlager JB, Yamabayashi Y, Franzen DL and Juneau RI 1989 *IEEE Photonics Technol. Lett.***1** 264–6

Schneider J 1995 *IEEE Photonics Technol. Lett.* **7** 354–6

Semenkoff M, Guibert M, Kerdiles JF and Sorel Y 1994 *Electron. Lett.* **30** 1411–13
Shi Y, Sejka M and Poulsen O 1995 *IEEE Photonics Technol. Lett.* **7** 290–2
Shimizu M, Yamada M, Horiguchi M and Sugita E 1990a *IEEE Photonics Technol. Lett.* **2** 43–5
Shimizu M 1990b *Electonics Lett.* **26** 1641
Sugawa T and Miyajima Y 1991 *IEEE Photonics Technol. Lett.* **3** 616–18
Tachibana M, Laming RI, Morkel PR and Payne DN 1991 *IEEE Photonics Technol. Lett.* **3** 118–20
Tesar A, Campbell J, Weber M, Weinzapfel C, Lin Y, Meissner H and Toratani H 1992 *Optical Materials* **1** 217–34
Urquhart P 1989 *Fibre Laser Sources and Amplifiers SPIE* **1171** 27–42
Whitley TJ 1995 *J. of Lightwave Technol.* **13** 744–60
Willner AE and Hwang SM 1995 *J. of Lightwave Technol.* **13** 802–16
Wu B and Chu PL 1994 *Opt. Commun.* **110** 545–8
Wysocki PF, Simpson JR and Lee D 1994 *IEEE Photonics Technol. Lett.* **6** 1098–100
Yamada M, Shimizu M, Horiguchi M, Okayasu M and Sugita E 1990a *IEEE Photonics Technol. Lett.* **2** 656–8
Yamada M, Shimizu M, Okayasu M, Takeshita T, Horiguchi M, Tachikawa Y and Sugita E 1990b *IEEE Photonics Technol. Lett.* **2** 205–7
Yamada M, Shimizu M, Takeshita T, Okayasu M, Horiguchi M, Uehara S and Sugita E 1989 *IEEE Photonics Technol. Lett.* **1** 422–24
Yang X, Zhang M and Yin G 1989 *Inter. J. Opto Elect.* **4** 397–403
Zemon S, Pedersen B, Lambert G, Miniscalco WJ, Andrews LJ, Davies RW and Wei T 1991 *IEEE Photonics Technol. Lett.* **3** 621–4
Zemon S, Pedersen B, Lambert G, Miniscalco WJ, Hall BT, Folweiler RC, Thompson BA and Andrews LJ 1992 *IEEE Photonics Technol. Lett.* **4** 244–7
Zervas MN, Laming RI, Townsend JE and Payne DN 1992 *IEEE Photonics Technol. Lett.* **4** 1342–4
Zervas MN and Laming RI 1995 *J. of Lightwave Technol.* **13** 721–31
Zhang J and Lit JW 1994 *IEEE Photonics Technol. Lett.* **6** 588–90
Zyskind JL, Giles CR, Desurvive E and Simpson JR 1989 *IEEE Photonics Technol. Lett.* **1** 428–30

EXERCISES

6.1 A fibre laser is made using two directional couplers and two fibre ring reflectors following the scheme Figure 6.10(d). The distance between the two couplers C is l_3. If l_i, T_i, R_i $(i = 1, 2)$ are the length, transmission, and reflection of each loop and α the attenuation in the fibre, establish the ratio I_0/I_i of the output signal power from the loop 2 to the input power of the loop 1.

6.2 Calculate the equations 6.5 and 6.6 from the equations 6.4.

Chapter 7

RECENT DEVELOPMENTS AND CONCLUSION

7.1 SYNTHESIS OF DEVICES DESCRIBED IN THIS BOOK

The purpose of this book is to provide a review of the main optical fibre devices and a reference book for work in photonics and lightwave. It reflects the substantial progress made in the area of optical fibre devices and provides a systematic description of linear and non-linear fibre devices. Following a survey of the physical processes exploited, the principles, manufacturing, properties, characterization, performance and applications have been examined. Optical fibre devices used in telecommunications and sensors have been described. Attention was devoted to fabrication and practical applications of these devices. The book covers components from connectors to more sophisticated systems using polarized light, solitons, amplifiers and lasers.

In the field of coupling, the devices described are connectors between fibres, and connectors or coupling systems between fibres and waveguides. Today, industrial connectors are fabricated in large numbers, mainly for optical communications but also for other applications as sensors. Pigtailed systems of good performance (laser diode and LED) are also available on the market. However particular systems need studies and development.

The first aim of this final chapter is to summarize the properties of all the devices that have been described in this book. Table 7.1 provides the summary.

The couplers (X, Y or star couplers) also available are essentially based on fused technology. However, couplers made in integrated optics also give good results and have the advantage of being fabricated in great numbers on a wafer; but they need specific coupling. Research

Table 7.1 Linear and non-linear optical devices

Fibre devices	Type	Physical principle	Fabrication
Tapers (expanders, concentrators)	MM SM	expansion of spot size	stretched fibre with fused spherical micro lens
Connectors	MM SM SM	coupling fibre to fibre to waveguide	butt coupling with micro-components V-grooves
Directional couplers		evanescent wave	polished fibre with an index-matching liquid
Polar independent X, Y couplers	MM SM	direct wave coupling	twisted and fused
$n \times n$ couplers	MM SM	micro-optic coupling	with cylindrical GRIN rod lens or silica microprisms
		electromechanical	MEMS–MOEMS
Polar dependent X-maintaining X-splitting	PMSM PMSM		
Bragg gratings	SM		holographic method
Wavelength filters	SM	mode leakage by bending	polished single-mode fibre with immersion liquid
		coupling from the core to a high index waveguide	as above, plus silica superstrate
		direct wave coupling by phase matching	tapered coaxial couplers
		wavelength selection	grating etched into polished fibre
Frequency shifters	SM PMSM	evanescent coupling between LP_{01} and LP_{11} coupling two polarizations of a birefringent fibre	periodic micro-bending on fibre (by squeezing) or acoustic wave flexure of long gratings microbends
Phase shifters	SM	Thermo-optic effect or stress	Bragg gratings

Table 7.1 *Continued*

Polarizers and polarization state controllers	PMSM	evanescent coupling of one polarization state	with birefringent crystal, thick metal film, etched fibre or D-shaped metal coated fibre
		bending effect (index change)	ordinary fibre coil (1/4 or 1/2 plate) or birefringent fibre (linear-elliptic)
		Faraday effect	birefringent twisted fibre with a solenoid
Isolators	MM SM	Faraday rotation	with crystal or plastic optical fibre surrounded by permanent magnets
Switches*	MM SM	radiation coupling (thickness variation piezo-electric translator or index variation)	high index waveguide sandwiched between two polished fibres coupler
		Bragg diffraction	acoustic transducer on the fibre
		optical Kerr effect (index variation)	dual core fibre
	SM	mechanical systems	MEMS–MOEMS taps
Wavelength multi-demultiplexers	MM SM	combination of various wavelengths channels	directional coupler or filters with external components Bragg gratings mechanical devices
Loops and rings–resonators–delay lines–circulators	SM		using X-couplers
Reflectors	SM	reflectance in a loop	fused taper coupler and loop
Lasers and amplifiers*	SM PMSM	Brillouin	silica fibre
	SM PMSM	Raman	silica fibre with dopants, e.g. Ge, P, H_2 to modify Raman gain

Table 7.1 *Continued*

	SM	four-wave mixing soliton effect	
	SM	fluorescence	single-mode rare-earth Nd^{3+} or Er^{3+} doped fibre
	REDF	spectrum	
Interferometers*	SM	interference between two waves	two single-mode fibres in the two arms or one arm with two polarizations
Polarimeters	PMSM		
Phase modulators*	SM	photo-elastic effect (see linear switch)	fibre sheathed with copolymer
Intensity modulators	SM		Acoustic wave

MM   Multimode fibre
SM   Single-Mode fibre
PMSM Polarization-Maintaining Single-Mode fibre
SMREDF Single-Mode Rare-Earth Doped Fibre

* Linear and non-linear devices (other devices are only linear)

efforts are still necessary to realize systems such as commutators or to connect N to N fibres (star couplers) or switches.

Devices using Bragg gratings have an important role in research and development, and have promising applications in telecommunications, sensors and signal processing.

Wavelength multiplexers and demultiplexers, frequency and phase shifters, wavelength filters, loops and rings, modulators and switches are key components for telecommunications.

All devices based on polarized light have great potential for development in the control of optimum propagation.

The studies of non-linearities in fibre stimulated Raman scattering, stimulated Brillouin scattering, parametric four-wave mixing and Kerr nonlinearities have permitted a better knowledge of wavefront propagation, and have led to new devices such as switches, amplifiers and to soliton propagation.

The fastest development in the field of fibre devices is certainly the recent discovery of rare-earth doped fibres and their application to the fabricating of amplifiers and lasers, which are now available on the market.

7.2 NEW DEVELOPMENTS

The other aim of this final chapter is to indicate applications in which optical systems have real and identifiable advantages and to identify some important technological and conceptual advances, which are likely to take place in the near future.

The applications of the optical fibre devices are today mainly for telecommunications, but probably in the future for sensors and signal processing.

The dense wavelength division multiplexing (WDM) optical networks, especially in a reconfigurable network, demand optical devices such as wavelength add/drop multiplexers and optical cross-connect (OXC) switches. Significant efforts have been devoted to the design of high capacity, flexible, reliable and transparent multi-wavelength optical networks.

Ultrahigh-speed all-optical time division multiplexed (TDM) systems are of great interest, and a single wavelength TDM system with a bit rate as high as 400 Gbit/s has been demonstrated. Optical switching devices with ultrafast responses are required in order to perform multi/demultiplexing in such high-speed optical networks. Non-linear optical loop mirrors (NOLM) and non-linear amplifying loop mirrors (NALM) are simple fibre-based devices, easy to produce. These devices possess optical switching properties when they are unbalanced by an asymmetric coupler and gain element. The NOLM has been one of the most successful devices for demonstrating a range of all optical processing: soliton switching, demultiplexing, wavelength conversion, and optical logic functions.

Fibre optics switches are used to reconfigure networks and increase their reliability. Low optical insertion loss and cross-talk are two of the most important requirements for optical switches. If opto-mechanical switches have lowest loss and crosstalk they are bulky, slow and expensive. Such devices based on fibres are better; however another attractive technology for making optical systems is micro-electro-mechanical systems (MEMS) technology, which can be coupled to optical fibre in order to construct new devices, for example, switches with N ports. It is possible to fabricate movable structures, micro-actuators, and micro-optical components, using low cost batch fabrication techniques similar to semiconductor electronic chip production processes. They can be monolithically integrated. Multi-wavelength optical switches for WDM applications have been described, using micro mirrors or for cross connections.

The acousto-optic effect in single-mode fibres has many applications in coupler and taper technology. Acousto-optic Bragg diffraction is an interaction between two or more light waves and an acoustic (mechanical) wave in optical material. An AO device can be made to act as an optical switch by turning the acoustic wave on or off and as a modulator or variable beamsplitter by changing the acoustic amplitude. It is also a frequency shifter because the index modulation travels along the acoustic wave leading to a Doppler shift.

Gratings, in particular long period fibre gratings (LPFGs), are important in a wide variety of telecommunications and sensing applications (Kashyap 1999). The development of Bragg grating applications in the last ten years is very important in research and industry. LPFGs are attractive fibre optic devices for use as band rejection filters with compactness, low insertion loss and low backreflection. They are of particular interest as gain-flattening filters for EDFAs, narrow band filtering, dispersion compensation, spectral shaping of broadband sources, optical fibre polarizers, laser stabilization and as optical sensors for measuring temperature, pressure or refractive index of liquids.

EDFAs operate in the optical fibre transmittance window at around 1550 nm and have obtained a wide acceptance in the optical fibre communication industry in the past few years due to compatibility with the standard silica fibres, low-loss coupling, low cross-talk, high gain and low noise figures (Digonnet 1993). Optically preamplified receivers employing EDFAs have now been implemented in multi-gigabit systems operating at more than 10 Gbit s^{-1}.

Multiple-wavelength fibre lasers are of interest for various applications such as WDM transmittance systems, optical sensing or optical signal processing. Simultaneous lasing at eight wavelengths has been achieved with an zirconium-doped fibre laser and several Bragg gratings written on the same segment of the fibre core (Wei et al. 2000). Other studies report the generation of high energy subpicosecond optical pulses.

In the future, new components may be obtained using recent technology or results such as holey fibre (Bennett et al. 1999) or photonic crystal fibres (Birks et al. 1997, Ortigosa-blanche et al. 2000, Ferrando et al. 2000, Randa et al. 2000). Photonic crystal fibres are guiding structures that possess particular properties that are unusual in conventional fibres. These are thin silica glass fibres that have a regular array of microscopic holes that extend along the whole fibre length. They can potentially be highly birefringent. The single-mode properties and mode intensity distribution differ from the standard step index fibre. A large mode range is possible, and these fibres are forecast to be good candidates for making fibre lasers with Bragg gratings (Sondergaard 2000).

They are also optical fibres with distributed uniform (or nonuniform) cores, semiconductor cylinder fibres (Kornreich et al. 1996) and infrared carrying fibres (1–6 μm). Fibres fabricated with new materials such as polymer are also promising for broadband optical amplification or in local area networks (LANs) (Peng et al. 1996, Tagaya et al. 1993, Ishigure et al. 2000).

As shown in this book, a great many devices have been conceived, characterized and constructed. But a large proportion of these devices, though they may have considerable potential applications, have not yet been developed commercially. The main reasons are technical problems involved in their realization and the high costs involved.

The optical fibre component world is expanding fast, and it is likely that the available tools for a system designer will increase significantly in the next decade.

REFERENCES FOR CHAPTER 7

Bennet PJ, Monro TM and Richardson DJ 1999 *Opt. Lett.* **24** 1203–5
Birks TA, Knight JC and P St J Russell 1997 *Opt. Lett.* **22** 961
Digonnet MFJ 1993 *Rare-earth doped fiber lasers and amplifiers* (Marcel Dekker: New York)
Ferrando A, Silvestre E, Miret JJ, Andres P and Andrès MV 2000 *Opt. Lett.* **25** 1328–30
Ishigure T, Koike Y and Fleming JW 2000 *J. Lightwave Technol.* **18** 178–84
Kashyap RK 1999 *Fibre Bragg Gratings* (Academic Press: New York)
Kornreich P, Cheng NS, Wu LM, Tung JT, Boncek R, Krol M, Stacy J and Donkor E 1996 *J. Lightwave Technol.* **14** 1694–6
Kishi N, Tayama K and Yamashita E 1996 *J. Lightwave Technol.* **14** 1794–800
Ortigoso-Blanch A, Knight JC, Wadsworth W J, Arriaga J, Mangau BJ, Birks A and PSt J Russell 2000 *Opt. Lett.* **25** 1325–7
Peng GD, Chu P, Xiang Z, Whitbread TW and Chaplin RP 1996 *J. Lightwave Technol.* **14** 2215–23
Randa JK, Windeler RS and Stenz AJ 2000 *Opt. Lett.* **25** 25
Sondergaard T 2000 *J. Lightwave Technol.* **18** 589–97
Tagaya A, Koike Y, Kinoshita T, Nihei E, Yamamoto T and Sasaki K 1993 *Appl. Phys. Lett.* **63** 883
Wei D, Li T, Zhao Y and Jian S 2000 *Opt. Lett.* **25** 1150–2

LIST OF SYMBOLS

Only notation that appears in several chapters is listed

a	Core radius
Ae	Effective area
AOD	Acousto Optic Device
b	Fibre radius (core+cladding)
B	Birefringence
c	Light velocity in empty space
C	Coupling coefficient
C'	Heat capacity
CR	Coupling ratio
CW	Continuous wave
d	Fibre core axes lateral offset at a joint
$D(\lambda)$	Group velocity dispersion $= \lambda^2 \partial^2 n / \partial \lambda^2$
DFB	Distributed feedback laser
E	Electrical field of the optical wave
EDFA	Erbium doped fibre amplifiers
FDM	Frequency division multiplexing
FP	Fabry-Pérot
FPM	Four photon mixing
FWHM	Full width at half maximum
FWM	Four wave mixing
g	Gain coefficient
G	Gain in dB
GVD	Group velocity dispersion $= D(\lambda)$
$h, \hbar$	Planck's constant, Planck's constant $\div 2\pi$
H	Magnetic field
HE_{11}	Fundamental mode (or LP_{01})
i	$(-1)^{1/2}$
Jn(.)	Bessel function of order n
k_0	Free-space propagation constant of the light
Kn(.)	Modified Bessel function of order n
L	Fibre length
LAN	Local area network
L_b	Beat length
L_c	Coupling length
LD	Laser diode
L_e	Effective fibre length
LED	Light emitting diode
LP_{01}	Fundamental mode (or HE_{11})
LP_{11}	Second order mode

List of symbols

MD	Multi demultiplexer
n	Average refractive index
n_e	Effective index
n^*	Non-linear index
n_1	Maximum core index
n_2	Inner cladding refractive index
n_3	Outer cladding refractive index
NA	Numerical aperture
NDS	Non-dispersion shifted
NLSE	Non-linear Schrödinger equation
P	Power
PC lens	Plano convex lens
Q	Non-linear coefficient
r	Radial distance from fibre axis
R	Fibre radius curvature
REDFA	Rare-earth doped fibre amplifier
rms	Root mean square
SBS	Stimulated Brillouin scattering
S_i	Stokes parameters
SOP	State of polarization
SPM	Self phase modulation
SRS	Stimulated Raman scattering
t	Time
T	Temperature
TDM	Time division multiplexing
T_r	Taper ratio
u	Normalized transverse propagation constant in the core
V	Normalized frequency
w	Normalized transverse propagation constant in the cladding
WDM	Wavelength division multiplexing
z	Abscissa along fibre axis
α	Fibre loss per unit length
α_s	Rayleigh scattering loss per unit length
β	Propagation constant
Δ	Core-index cladding relative index difference
ε	electric permittivity
η	coupling efficiency
λ	Light wavelength in material
λ_c	Cut-off wavelength of a mode
λ_0	Light wavelength in empty space
Λ_B	Bragg period
μ	Magnetic permeability
ν	Optical frequency $\nu = \omega/2\pi$

v_a	Acoustic frequency
θ_{ca}	Critical angle in the fibre
τ	Delay time
φ	Azimuth in cylindrical coordinates
ϕ	Total phase of the transmitted field
χ	Susceptibility
ψ	Field amplitude
ω	Optical pulsation $\omega = 2\pi v$
ω_0	Mode field radius of the LP_{01} mode
Ω	Angular velocity

SOLUTIONS TO EXERCISES

Solutions to Chapter 1

1.1 $R = [(n_1 - n_0)/(n_1 + n_0)]^2 = 0.035$
$T = 1 - R = 0.965$

1.2 **a** $\Delta = (n_1^2 - n_2^2)/2n_1^2 = 0.0065$
b $NA = n_1(2\Delta)^{1/2} = 0.1752$
c $\sin\theta_{ca} = NA/n_0 = 0.1738 \quad \theta_{ca} = 10°$
d $\delta L = Ln_1\Delta/n_2 = 6.5$m and $\delta t = Ln_1/c = 33$ ns

1.3 **a** $P(z) = P_0 10^{-\alpha L/10} = 0.63$ mW
b $P(z) = 10^{-3}\, mW = 10^{-\alpha z}$ and $z = 65.1$ km

1.4 $V = ak NA = akn_1(2\Delta)^{1/2} \quad \lambda = 1.549\,\mu$m
$n_1 = 1.414 \quad n_2 = 1.412$

1.5 $\nabla \wedge (\nabla \wedge \mathbf{H}) - \nabla^2 \mathbf{H} = k^2 \varepsilon_m \mathbf{H}$

Solutions to Chapter 2

2.1 $A = 2\Delta n/(na^2) \quad Z = 2\pi(A)^{-1/2} = 7$ mm
$Z = \pi(4A)^{-1/2} = 1.75$ mm

2.2 $W = 4\pi I_0 \int r\, dr \int \sin\theta\, \cos\theta\, d\theta = \pi I_0 \, a^2 \sin^2\theta_m$
$\sin\theta_{ca} = NA = n_1\sin\theta_c$ and $I_0 = 0.16$ GW/m^2

2.3 $\eta = W_o/W_i = 1 - [4z\, NA/3\pi\, a\, n_0]$
$\eta = 48.77$ mW

2.4 $\eta = 1 - 2\rho/\pi$ and $\rho = d/a$
$\eta = 0.34$ dB

2.5 **a** $W_o = I_0\pi^2 b^2$ and $I_0 = 0.317$ MW/m^2
b $W_i = I_0\pi^2 a^2\, NA^2 = 0.09$ mW
c $A_f < A_s$ and $\Lambda = W_i/W_o = a^2 \sin^2\theta_{ca}/b^2 = 0.017$
$W_i = 0.09$ mW

2.6 Between $z = 0$ and $z_M = a/\text{tg}\,\theta_{ca} \# a\,/\,NA$ and $z_M = 159\,\mu$m

Solutions to Chapter 3

3.1 $F = 1$ (2 identical fibres) $\quad\quad C = \pi/L_b = 628.3 \times 10^3$ m^{-1}
$P_1 = P_0 \cos^2(\pi/L_b)\,z$ and $\quad P_2 = P_0 \sin^2(\pi/L_b)\,z$
$z_1 = 250\,\mu$m $\quad\quad P_1 = 0 \quad\quad P_2 = 100$ mW
$z_2 = 2.50$ mm $\quad\quad P_1 = 100$ mW $\quad P_2 = 0$

Solutions to exercises

3.2 $\varphi = 2\pi\delta/\lambda = 2\pi n L/\lambda$
$\delta\varphi = k L (\partial n/\partial T + n\alpha)\delta T = 47.7$ radians

3.3 $n_1 \sin\alpha_m - n_e = m\lambda_B/\Lambda \quad n_i = n_1 \quad \alpha_m = -\pi/2 \quad m = -1$
$\lambda_B = 2n_e\Lambda = 1.55$ μm
Every modification upon n_1 or Λ induces a variation $\delta\lambda_B/\lambda_B$

3.4 $\lambda_0 < n_e\Lambda \quad \sin\alpha_m = 1 + m\lambda_0/n_e\Lambda$
$\alpha_0 = \pi/2 \quad \alpha_{-1} = 46°55 \quad \alpha_{-2} = 25°15 \quad \alpha_{-3} = 10°26$
$\alpha_{-4} = -5°5 \quad \alpha_{-5} = -21°7 \quad \alpha_{-6} = -39°79 \quad \alpha_{-7} = -66°63$

3.5 During the delay of propagation upon a coil, the beam splitter turns of $\Delta l = R\Omega\tau$ with $\tau = 2\pi R/v$ (v velocity of light in silica). The length difference between the two propagations is $\Delta L = 4\pi R^2\Omega/v$ and for N coils $\Delta\varphi = kn\Delta L$.
So $\Delta\varphi = 8\pi^2 R^2 N\Omega n^2/\lambda_0 c = 2.8 \times 10^{-4}$ rd.

Solutions to Chapter 4

4.1 $\Delta n_a = Sn^3 kF/4b = 5.5 \; 10^{-4}$
$\Delta\varphi = 2\pi\Delta n_a L/\lambda = 239$ rd

4.2 $\Delta n_i = 2\pi/kL_b = \lambda/L_b \quad L_b = 500$ μm
$\Delta n_i = 3.1 \times 10^{-3} \quad\quad \theta = \Delta n_a/\Delta n_i = 0.177$ rd $= 177$ mrd

4.3 $\delta n = d (b/R)^2 \quad\quad \delta\varphi = k\delta nL = kdb^2 N 2\pi/R = 2\pi/m$
$R = 2\pi db^2 mN/\lambda$

4.4 a $\varphi = 2\pi\delta nL/\lambda = 2\pi/m = \pi/2$ and $m = 4$
b $\delta n = 0.133 (b/R)^2$ and $\delta n = \lambda/4L \quad L = 2\pi RN$ so
$R = 0.133 (8\pi Nb^2/\lambda) \quad R = 37$ mm

Solutions to Chapter 5

5.1 –a $P(L) = P(0)$ and $K = 1, L_e = 1/\alpha_p$
$L_e = g_r P(0)/\alpha_p \alpha_s A_e$
–b $G = 21.6 dB$

5.2 $-\nu_B = -17.3/\lambda_P = 13.3$ GHz
$\nu_s = \nu_B[c/(nV_a) - 1]/2 = 2.29 \times 10^{14}$ Hz
$\nu_s = \nu_p - \nu_B = 2.29 \times 10^{14}$ Hz

5.3 $\nu_B = 11.61$ GHz
$\Delta\nu_B = 17.6$ MHz
$g_B = 4.2 \times 10^{-11}$ mW^{-1}

5.4 $D_c = -\dfrac{\lambda}{c}\dfrac{d^2 n}{d\lambda^2}$ and $D(\lambda) = -\lambda^2 \dfrac{d^2 n_e}{d\lambda^2}$

$D(\lambda) = \lambda c D_c = 1.32 \times 10^{-6} * 3 \times 10^8 * 1 \times 10^{-12}/(10^{-9} 10^3)$
$D(\lambda) = 4 \times 10^{-4}$
$\Delta k_m = 2\pi \lambda D(\lambda) \Omega^2$ with $\Omega = \dfrac{|\omega_p - \omega_s|}{2\pi c} = \dfrac{|\lambda_s - \lambda_p|}{\lambda_s \lambda_p}$ (cm^{-1})
$\Omega = (0.35 \times 10^{-6})/(1.32 * 1.67 \times 10^{-12}) = 1580$ cm^{-1}
$\Delta k_m = 2\pi * 1.32 \times 10^{-6} * 4 \times 10^{-4} * 1580/10^{-2} = 517 \times 10^{-6}$ m^{-1}

5.5 $z = 0.322 \left(\dfrac{\pi^2 c}{\lambda_0^2}\right) \dfrac{\tau^2}{|D|}$

$z = 3.91 \times 10^{26} \tau^2 = $ 977 kms for $\tau = 50$ ps
$\phantom{z = 3.91 \times 10^{26} \tau^2 = }$ 352 kms for $\tau = 30$ ps

Solutions to Chapter 6

6.1 Inside a resonator the field is $E = t + tr^2 \exp(-i\varphi) + tr^4 \exp(-2i\varphi) + \cdots$
The sum is $E = t/[1 - r^2 \exp(-i\varphi)]$
In the case of the two loops, the phase $\varphi = \beta(l_1 + l_2 + 2l_3)$ and the intensity is given by $I = E E^* = t^2/[1 + R_0^2 - 2R_0 \cos \varphi]$ with $R_0 = R_1 R_2$
We have $1 + R_0^2 - 2R_0 \cos \varphi = 1 + (R_1 R_2)^2 - 2R_1 R_2 \cos \varphi = 1 + (R_1 R_2)^2 - 2R_1 R_2 + 2R_1 R_2 - 2R_1 R_2 \cos \varphi = (1 - R_1 R_2)^2 + 2R_1 R_2 (1 - \cos \varphi)$
The transmitted power is $t^2 = T_1 T_2 \exp(-2\alpha l_3)$
So $I_0 = I_i T_1 T_2 \exp(-2\alpha l_3)/[R^2 + 4R_1 R_2 \sin^2 \varphi/2]$ with $\varphi/2 = B$ and $R = 1 - R_1 R_2$

6.2 The equations (6.5) and (6.6) are obtained using $N_4 = 0$ (without ESA)
$-dN_1/dt = dN_2/dt + dN_3/dt = 0$ (steady state)
$N_3 \approx 0 \qquad N_1 + N_2 = N$
The first equation of (6.4) gives (6.5)
The two following equations of (6.4) gives (6.6)

Index

Amplification, amplifiers
  based on stimulated Brillouin scattering, 179–186
  based on stimulated Raman scattering, 172–179
  using rare earth doped fibres, 25, 222–238
Attenuation, 16, 145

Birefringence, 19–22, 129–134, 155
Birefringent fibres, 19, 105, 202–209
Bragg gratings, 93–99, 101
Brillouin scattering, 179–180
  Brillouin amplifier, gain coefficient, 180–183
  Brillouin laser, 183–186
Butt end coupling, 31–33, 60, 61, 66, 67

Circulators, 117
Connectors, 58
Couplers
  birefringent fibre polarization couplers, 137–140
  non-linear couplers, 196, 199–202
  polarization maintaining, 134
  polarization splitting, 135
  star couplers, 90–93

X couplers, 80–89, 134–137, 241
Y couplers, 89
Coupling
  from fibre to fibre, 49–58
  from fibre to waveguide, 58–64
  from semi conductor laser or LED into fibre, 64–69
  theory for circular fibres, 76–80
Cut-fibre method, 17

Delay lines, 117
Depolarizers, 151

Effective index, 14
Expansion beam component (*see also* tapers), 39–49

Faraday effect, 133
Fibre, 1–10
  birefringent fibres, 19–20, 130, 202–209
  characterization, 15–18
  doped fibres, 25, 220–225, 235 (*see also* rare earth doped fibre/amplifiers/lasers)
  fabrication, 22–25
  material, 24
  multimode, 6–14, 50–53

267

268 Index

panda, 19, 130, 134, 210
single mode, 14–15
Filters
   modal filters, 110
   wavelength filters, 106–110

Gratings, 57, 93–99, 101, 239–241
GRIN lens, 33–35, 54–56, 67, 68
Grooves, 61, 64
Group velocity dispersion, 187, 192

Index (refractive), 2–5, 41, 165,
   non-linear index, 165–166, 189
Index-matching fluid, 32
Interferometers, 118–122, 160–162
   Fabry-Pérot, 238
   Mach-Zehnder, 102, 103, 118, 119, 121, 152
   non-linear interferometer, 211
   Sagnac ,120
Isolators, 159

Kerr non-linearities, 165–167, 188–191

Lasers
   Brillouin lasers, 183–186
   diode, 26–27, 64–69
   rare earth doped fibre laser (REDFL), 238–247
   soliton laser, 195
LED, 28, 64–69
Light sources, 25–28,
   superfluorescent, 247
Logic gate, 212
Loops, 116, 177, 183 , 210, 242–243
Losses
   alignment losses, 49
   attenuation, 16–17

Material dispersion, 8
Maxwell equations, 10–14
Microcomponents, 33–39, 54–58
Microlenses, 35–39
Mirrors, 210, 242–243
Mode locking fibre laser, 245
Modes, 10–15
Modulators, 115–116, 212–214
   intensity modulators, 116
   phase modulators, 115
Multiplexers, 99–103

Non-linearity, 165–167
Normalized frequency, 14, 87
Numerical aperture, 8, 15, 26

Optical pulse compression, 191

Parametric four wave mixing, 186
Poincaré sphere, 21, 202–203
Polarimeters, 160–162
Polarization, 129, 188
Polarization state controllers, 153–159
Polarizers, 140–151
   circular polarizers, 151
Preform, 23–24
Propagation in optical fibres, 10–15

Q–switching, 244

Raman scattering, Raman effect , 167, 172
   amplification, Raman gain, 172–179
Rare earth doped fibres, 25, 221
   amplifiers 222–225, 236–238
   Er^{3+} amplifiers, 225–234
   Nd^{3+} amplifiers, 223, 235
   other REDFA Pr^{3+} Yb^{3+}, 235

REDFLaser, 238–247
Reflection, 2
Reflectors, 243
Refraction, 2
Resonators, 117, 183–186, 242
Retro-diffusion method, 18, 32
Rings, 116, 117, 183–186

Schrödinger
   non-linear Schrödinger equation, 196–199
SELFOC *see* GRIN lenses
Shifters
   frequency shifters, 103–105
   phase shifters, 105
Solitons, 186, 191–199
Splice, 49–54
   circular fields, 49
   elliptic fields, 53

multimodes fibres, 50
single mode, 52
Spot size, 41, 44
Stokes matrix, 132–133
Stokes parameters, 20
Switches
   linear switches, 110–114
   soliton switching, 196
   switches using birefringent fibres, 202
   switches using non-linear couplers, 199
   switches using non-linear fibre loop mirror, 210

Tapers, 39–49, 56–58
Taps, 114–115
Time delay, 8
Transistor (optical fibre), 214
Tunable laser, 246–247